CW01401839

THE NORTHERN

SOVIET EXPLOITA___ __ ___
NORTH EAST PASSAGE

SCOTT POLAR RESEARCH INSTITUTE

SPECIAL PUBLICATION NUMBER 1

THE
NORTHERN SEA ROUTE

SOVIET EXPLOITATION OF THE NORTH EAST PASSAGE

BY

TERENCE ARMSTRONG

PUBLISHED FOR THE

SCOTT POLAR RESEARCH INSTITUTE

CAMBRIDGE

AT THE UNIVERSITY PRESS

1952

CAMBRIDGE UNIVERSITY PRESS
Cambridge, New York, Melbourne, Madrid, Cape Town,
Singapore, São Paulo, Delhi, Tokyo, Mexico City

Cambridge University Press
The Edinburgh Building, Cambridge CB2 8RU, UK

Published in the United States of America by Cambridge University Press, New York

www.cambridge.org
Information on this title: www.cambridge.org/9780521232630

First published 1952
First paperback edition 2011

A catalogue record for this publication is available from the British Library

ISBN 978-0-521-23263-0 Paperback

NOTE

This volume was the first publication of
the Scott Polar Research Institute.

CONTENTS

MAPS

ILLUSTRATIONS

PREFACE

The development of a freight-carrying sea route along the north coast of Siberia has been the subject of boasts inside the Soviet Union and scepticism outside it. Considerable importance attaches to its success or failure. If successful, there will be far-reaching effects in the spheres of economics and strategy. Whether a success or a failure, countries interested in polar navigation may be able to learn from Soviet experience.

Yet there has been little written in English about the Northern Sea Route, as it has come to be called. By far the best account, and the only one to make anything like full use of the sources available, is to be found in T. A. Taracouzio's *Soviets in the Arctic* (New York, 1938). This treats a much wider subject than the development of the Northern Sea Route, however, being concerned also with economic and political problems over a large area of mainland. Apart from this there are only short sketches. The best of these is Professor Kenneth Mason's "Notes on the Northern Sea Route" (*Geographical Journal*, Vol. 96, No. 1, 1940, p. 27–41), which is a very good digest of the information available at that time, but is based almost entirely on English language sources. There are several travel books by people who made voyages on parts of the route: Leonard Matters's *Through the Kara Sea* (London, 1932), Bosworth Goldman's *Red road through Asia* (London, 1934), Ruth Gruber's *I went to the Soviet Arctic* (New York, 1939) and H. P. Smolka's *Forty Thousand against the Arctic* (London, 1937). These are of undoubted interest, but contain little reliable information on the problem as a whole, however accurate may be the accounts of what their authors themselves saw.

Even in Russian no general works on the subject are available. Mention should be made however of V. ·Yu. Vize's fine history of exploration *Morya sovetskoy arktiki* [*Seas of the Soviet Arctic*] (Leningrad, 1936), which provides the main outline of the historical background. The 1948 edition of this work, revised and augmented, has just become available in this country.

It is the aim of the present study, which is based almost entirely on Russian language sources, to provide a fuller account than any of those mentioned, and to bring the story up to date. The source material consists chiefly of Soviet books and journals published during the last thirty years. The Soviet Union produces a great number of scientific and technical journals, and Arctic studies claim their proportion of these. The standard of accuracy of the papers contained in them is not always high. Statistics in particular are suspect, not so much because they may have been deliberately distorted but because of slipshod collecting of data and imprecise presentation. Nevertheless the journals contain a great deal of information. But the difficulty for the student working outside the Soviet Union is to find this literature. Not only the various periodicals on a given subject, but the issues of a single periodical are scattered among widely separated libraries. In time fairly complete sets of most journals can be pieced together, but even when the resources of this country, the rest of Western Europe and North America have been tapped, there remains an almost complete blank for the post-war period. The withholding of certain information has clearly become a matter of policy. The gap is serious since the events of the last few years would provide a good measure of the success or failure of the intensive work done in the 1930's. But since there seems at present to be no immediate likelihood of a change of policy which would make the information available, it has been thought worth while to take the story as far forward as possible and estimate the value of the whole venture on the evidence available.

The Northern Sea Route is defined, for the purpose of this book, as a system of shipping lanes traversing the coastal waters north of Siberia, bounded by Bering Strait in the east and by the straits between the Barents Sea and the Kara Sea in the west (see map 9 at end of volume). But these limits are not rigid; for instance at times it will be necessary to consider extensions at each end to the terminal ports which are outside the confines of the route proper. The limit of time is determined by the fact that the emphasis is upon the development of a freight route, as opposed to the discovery of a navigable waterway in that part of the world; and it was in the second half of the nineteenth century that freighting possibilities were first seriously considered. In order to split the period into manageable pieces it is convenient to divide it into two unequal lengths at the year 1932. Development up to that year may be looked upon as a historical introduction to the more important events that followed, and this will form the first part of this study. The second part will be devoted to consideration of various aspects of the complex pattern of development inaugurated in 1932 by the setting up of a special Government department in charge of Arctic affairs; and to an examination of the uses to which the route was put.

Transliteration of Russian names is by the system published by the United States Board on Geographic Names in 1944 and approved by the British Permanent Committee on Geographical Names in 1948 (Permanent Committee on Geographical Names for British Official Use, *Table for the transliteration of Russian geographical names*, London, December 1948). Place-names are given in transliteration of their full Russian form, and a glossary which includes geographical terms is to be found on p. 138.

Efforts have been made to ascertain the type of ton referred to in tonnage figures. In certain cases however neither the original source nor subsequent research makes the point clear, and here the word ton is left unqualified. Mileage is expressed in statute miles unless stated to the contrary.

The writer wishes to express his sincere thanks to the many individuals and institutions that have provided help and advice: particularly to Dr B. B. Roberts and Professor F. Debenham of Cambridge, and Mr Willis C. Armstrong of Washington, D.C.; to the Arctic Institute of North America and the Canadian Defence Research Board, whose financial aid helped to make possible extremely valuable visits to United States and Canadian libraries; to the Treasury Committee for Studentships in Foreign Languages and Cultures, whose award of a senior studentship enabled the work to be completed and the writer to visit libraries in Paris and Prague; and above all to the Bishop of Portsmouth, Dr G. C. L. Bertram and the staff of the Scott Polar Research Institute, Cambridge, within whose hospitable walls this study has been composed.

T.E.A.

January, 1952

PART I

THE NORTHERN SEA ROUTE BEFORE 1932

1. THE KARA SEA ROUTE TO THE OB' AND YENISEY

The earliest voyages in the Kara Sea were made by Russians, probably in the first half of the sixteenth century (21).[1] Traders or hunters from Kholmogora, close to modern Arkhangel'sk, sailed out of the White Sea or the river Pechora to the rivers Ob' and Taz. The route taken was normally through Yugorskiy Shar, the most southerly strait joining the Barents Sea and the Kara Sea; thence to Baydaratskaya Guba, and across the base of Poluostrov Yamal to the southern part of the Ob' estuary by way of a river system which allowed only a short portage across the watershed. This route was much used in the late sixteenth and early seventeenth centuries. From 1600 its eastern base was the settlement of Mangazeya on the river Taz, and from here it was not difficult to continue by inland waterways to the Yenisey. It is likely that voyages were made at about this time to points farther east than the Ob' estuary, even to the Laptev Sea. Evidence in support of this came to light recently when a number of objects which had evidently belonged to a shipwrecked wintering party were found on the east coast of Taymyr (118). In 1620 however a decree of Tsar Mikhail Fedorovich forbade trading by these routes. The reason for this interdiction was apparently fear that foreigners and Russian merchants would evade payment of dues at Arkhangel'sk by trading through the Kara Sea. The route to Mangazeya was soon forgotten. Although occasional hunting parties sailed into the Kara Sea after 1620, trading vessels did not return to these waters for over 250 years.

The early navigators to the Ob' and Yenisey left no written information in the form of charts or descriptions of the coast. The south-western littoral of the Kara Sea was therefore re-explored and described between 1734 and 1740 by the Russian Great Northern Expedition, a large and remarkable organisation to which the Russian Navy gave the task of describing the whole northern coast of the Empire from the Pechora to Bering Strait. This expedition was, in spite of all sorts of difficulties, astonishingly successful. The parties detailed to cover the Pechora-Ob' and Ob'-Yenisey sectors completed their work, but only with difficulty and after a number of attempts. Perhaps these difficulties explain why no attempt was made afterwards to reopen the route to the Ob' or Yenisey.

Between 1820 and 1839 nine Russian expeditions went to Novaya Zemlya, principally in order to map the coast and do hydrographic work. Bad ice conditions were not infrequently encountered in the Kara Sea off the east coast of Novaya Zemlya. This led two leaders of expeditions, Lieutenant F. P. Litke and Academician K. Ber, to make unfavourable comments on the Kara

[1] All references are listed on pp. 139–54.

Sea and the possibility of sailing in it. Ber in 1838 called it an "ice cellar", and this view of the Kara Sea was generally accepted for the next two decades. Another journey was made to the Kara Sea in 1860. In that year Lieutenant P. P. Kruzenshtern attempted to reach the west Siberian rivers; he left the Pechora on 10 September but was compelled to turn back, because of bad ice and the lateness of the season, when he had gone only a short way past Karskiye Vorota. In 1862 he was persuaded to try again. This time he was caught in the ice in Yugorskiy Shar. The ship drifted eastwards and finally

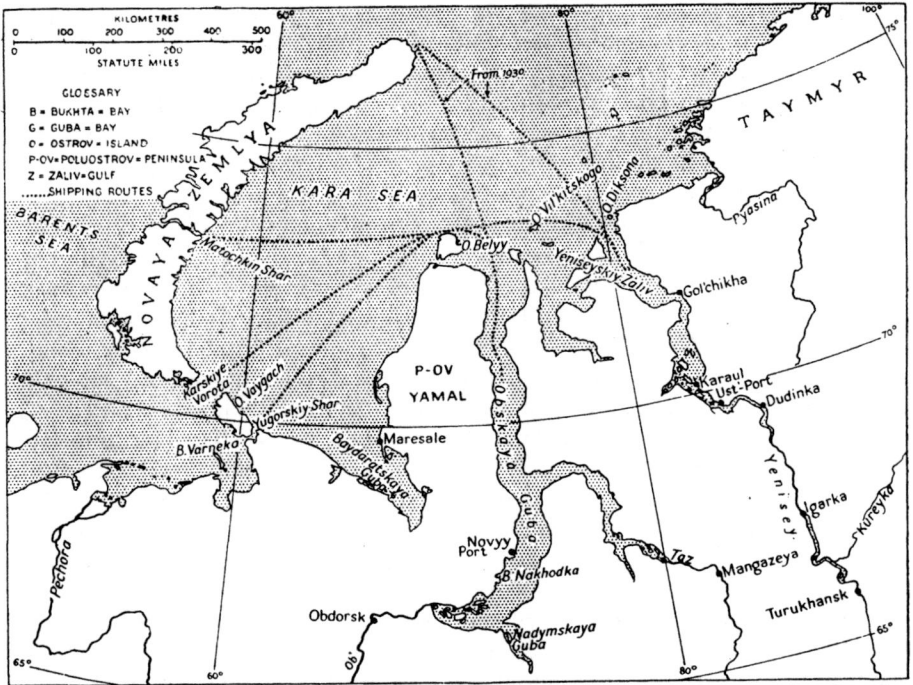

Map 1. The Kara Sea route

sank, and the crew later with difficulty reached Yamal. On his return Kruzenshtern wrote: "There can never be any sort of sea route through the Kara Sea to the mouths of the Ob' and Yenisey"(301). As might be expected, the effect of this voyage was to confirm in the public estimation the opinions of Ber and Litke.

Meanwhile, in the face of this unfavourable expert opinion, the desirability of finding a commercially usable waterway into and out of Siberia was being urged by a Russian merchant named M. K. Sidorov. Sidorov owned a large number of gold mines near the Yenisey and he wanted to import machinery for them. He also wanted to build up communications in general in that part of the world, a desire very natural in a business man who wants to develop a backward region. He was the first man to appreciate the significance of the rivers Ob' and Yenisey as a possible link between central Siberia and the out-side world, and he devoted most of the rest of his life to championship of this cause in particular, and of development of the Russian north in general. In 1859 he wrote a memorandum to the Governor of the Yenisey area "On the possibility of a sea route from Europe to east and west Siberia via the mouths of the rivers Ob' and Yenisey" (149). This met with no response. In 1862 as we have seen Kruzenshtern made his second attempt in the Kara Sea, and this attempt was made through Sidorov's initiative and at his expense. Soon afterwards Sidorov formed a company whose object was trade with Siberia by sea. He offered a prize of 14,000 roubles to the first man who should sail to the Yenisey, but no one responded. In 1869 he planned to attempt the voyage himself in the schooner *Georgiy*, but was delayed in starting and in the end had to drop the idea (355). All this time he had been trying to arouse interest in the north of Russia, and had published a number of articles on the subject. His book *Sever Rossii* [*The north of Russia*], published in 1870, was a collection of these writings. He had arranged exhibits of the natural wealth of north Russia at many international exhibitions in Russia, Germany and England. But none of these efforts had yet produced tangible results. His plans for the exploita-tion of Siberian raw materials, and the parallel development of traffic on the Ob' and Yenisey as the essential routes for import and export, remained only plans. The "ice cellar" legend was still too strong.

In the late 1860's however the Kara Sea was once again reached by ships from the west. Norwegian sealers found that there was good hunting to be had east of Yugorskiy Shar and Karskiye Vorota. The island of Vaygach was cir-cumnavigated by Elling Carlsen in 1868. In 1869 24 sealers went to the Kara Sea. Major John Palliser, an Englishman, sailed on a hunting trip to the neighbourhood of Ostrov Belyy in the same year. In each succeeding year Norwegian sealers entered the Kara Sea. It became clear that the Kara Sea, or at least parts of it, were navigable for a period in the summer after all.

It was not until 1874 that any serious attempt was made to reach the mouths of the Ob' or Yenisey. Captain Joseph Wiggins, a British seaman, had heard through Dr Petermann, the famous German geographer, that Sidorov was trying to realise the idea of a route to Siberia through the Kara Sea. Wiggins grew interested and came to the conclusion that the influence of the Gulf-stream would in fact make such a route navigable. He knew little or nothing of the Norwegian voyages, since the Norwegian skippers, not wishing to give away trade secrets, had kept quiet about their new hunting grounds. In June 1874 he set out in the *Diana*, a steamer of 103 tons net. He reached the Kara Sea at the end of June and had difficulty with ice, because it was still early in the season. At the end of July he reached the northern end of Obskaya Guba, east of Ostrov Belyy. At this point he unfortunately had to turn back: his food was getting low, his crew was not very keen, and the *Diana* drew twelve feet, too much in Wiggins's opinion to allow her to explore southwards in the shallow waters of Obskaya Guba. This voyage, in spite of its limited success, was none the less one of the landmarks in the pioneering of the Kara Sea route.

In the following year Wiggins again sailed to the Kara Sea, this time in a 27-ton fishing boat, the *Whim*. His object was to survey Baydaratskaya Guba in order to rediscover the sixteenth-century route across Yamal. But the *Whim* never reached the Kara Sea because too much time was wasted in collecting maps at Arkhangel'sk and the season broke early. A much more important voyage was made this year, however. Baron A. E. Nordenskiöld, a Swedish scientist who had already made several expeditions to Spitsbergen, decided to sail to the Yenisey and then up the river. This expedition was principally scientific, but Nordenskiöld was also aware of the commercial implications of a successful voyage to the Yenisey. He took his party of four scientists on a sealer of 70 tons, the *Pröven*. The voyage was entirely successful. The *Pröven* reached a small island north of the Yenisey estuary on 15 August. This island had a good harbour and Nordenskiöld called it Dicksons Havn after Oscar Dickson who had financed the expedition (the island is now called Dikson or Ostrov Diksona). Here Nordenskiöld and the scientists left the *Pröven*, which then returned to Norway, and they went up the river. The party at first used a small boat they had brought with them; then a little way upstream they transferred to a river steamer which took them right up to Yeniseysk. This voyage was the first known occasion on which anyone had crossed the Kara Sea to the mouth of the Yenisey. Nordenskiöld reports that there were two steamers on the river in 1875; both, he says, were really travelling stores owned by merchants and were built to carry neither passengers nor freight other than their owners' wares. Boats had in fact first sailed on the Yenisey a long time before, probably at the end of the sixteenth century. The boat which Nordenskiöld found had been plying between Gol'chikha and Yeniseysk since 1863 (264).

Sidorov meanwhile continued his energetic propaganda for the Kara Sea route. He insisted that an occasional voyage by foreigners was not enough: freight must be transported. He made another attempt to get a cargo taken, but the project again fell through. A fellow gold-mining magnate, A. M. Sibiryakov, grew interested in the idea; he was equally prepared to spend large sums in promoting it. Sibiryakov started by helping to finance Wiggins and Nordenskiöld, both of whom wanted to make another voyage in 1876. Nordenskiöld wanted to show that the success of the *Pröven* in the previous year was not simply the result of good weather. He chartered the *Ymer*, a steam freighter of 400 tons burden, and this time took on board some samples of Swedish manufactured goods. His voyage was successful. The goods were left at the mouth of the Yenisey and were taken southwards later in the season by river steamer. The *Ymer* returned to Norway. Wiggins, though a sailor first and last and in no way a business man, was always primarily interested in the benefits that his voyage would bring to commerce. He tried to interest a number of merchants in St Petersburg, but unsuccessfully. In the end money was put up by Sibiryakov and a rich English yachtsman named Charles Gardner. Wiggins bought the *Thames*, a steamer of 120 registered tons, and set sail with British samples from Sunderland. He did some further exploratory work in Baydaratskaya Guba, but failed again to find the waterway across

Yamal. He then tried to enter the Ob' estuary, but adverse winds and currents prevented him. So he entered the Yenisey. Sidorov had arranged for a cargo of graphite to wait for him at the mouth of the Yenisey; but Wiggins missed the schooner carrying this cargo and went on upstream to the river Kureyka, a tributary some 420 miles from the sea and the site of the graphite mines. Wiggins decided to leave the *Thames* at the Kureyka for the winter. The most important result of this voyage was the proof it provided that ocean-going ships could penetrate far up the Yenisey. The two collections of samples brought to the river in this season constitute the first commercial cargoes to cross the Kara Sea. Wiggins had some trouble at Turukhansk with local customs officials who tried to confiscate his samples—the first indication that difficulties were not to be expected only from the elements.

The successes of 1876 led to increased activity in the following year. The steamer *Luise* of 170 tons, commanded by Captain Dahl, sailed from Lübeck up the Ob' and thence the Irtysh to Tobol'sk. She was thus the first ship from the west to reach the Ob'. Unfortunately there is no detailed information on this voyage. Sibiryakov and Sidorov were both active again in 1877. Sibiryakov bought, in Nordenskiöld's name, the steamer *Fraser*. He loaded her with sugar, tobacco and some machinery for gold mining. Under the command of Captain Eduard Dallmann, who had navigated in Antarctic waters in 1873–4, she sailed from Bremen to the Yenisey and unloaded her cargo at the mouth of the river. A return cargo of grain which Sibiryakov intended to export to Europe was delayed on its way downstream by the machinations of merchants who disapproved of the idea of a Kara Sea route, and the *Fraser* had to sail without it. She made the exceptionally fast time of 8½ days from the Yenisey to Hammerfest; and this included a wait of two days on route. Meanwhile Wiggins was trying to get the *Thames* downstream from the Kureyka where she had successfully wintered. Unfortunately the ship went aground in the river and had to be abandoned. Wiggins wanted to continue in a 50 ft. lighter, the *Ibis*, which a travelling companion of his—Henry Seebohm, the ornithologist—had bought in Yeniseysk; but his crew refused to take such an unseaworthy craft beyond the river mouth. The *Ibis* was then bought by Captain Shvanenberg, a skipper acting for Sidorov. Rechristened *Utrennyaya Zarya*, she was sailed across the Kara Sea, round Scandinavia to St Petersburg with a cargo of graphite, fish and timber. This was the first cargo to leave the Yenisey by sea, and the occasion was something of a triumph for Sidorov. He had finally proved his point, and those who for nearly twenty years had dismissed his idea as absurd now congratulated him. But he could not take advantage of his success; his funds were exhausted.

Sidorov does not appear to have again taken an active part in furthering his project, though he continued to keep the cause of north Russian development as much in the public eye as possible by his exhibits at international exhibitions and by his publications. He was deeply disappointed with the reaction of his own government to his plans. He writes in 1882: "Looking back over my activity I must say with regret that during twenty years I have not had any co-operation. The administration has opposed me, although I asked for neither privileges nor help" (302). He saw clearly that if the Russian Government refused to allow its own nationals to develop the north, foreigners would quickly come in, do the pioneer work and reap the benefit. This thought disturbed him. He used to quote to obstructive officials the example of Peter the Great, whose policy was to give a free hand to anyone who wanted to exploit the natural resources of the country. He did succeed in organising navigation on the river Pechora, which flows into the Barents Sea. But most of his dreams remained unfulfilled in his lifetime. He died in 1887. By this time it is true there had been some fairly considerable exchanges of goods at the mouths of the Ob' and Yenisey; but the Russian Government was still taking no interest at all in what was happening.

The most important event of 1878 as far as navigation in these waters was concerned was the start of A. E. Nordenskiöld's attempt to sail the whole length of the North East Passage from Atlantic to Pacific in the *Vega*. This great voyage started extremely well, and the whole distance was very nearly covered in the summer of 1878. But the *Vega* unfortunately could not negotiate difficult ice when she was a short way from Bering Strait. She had to winter off the north coast of Chukotka and completed the voyage the following year. The significance of this voyage in the development of the Northern Sea Route as a whole is clear. Here however we are concerned with its relation to voyages to the Ob' and Yenisey. Nordenskiöld agreed to convey as far as the Yenisey two ships chartered by Sibiryakov: the steamer *Fraser* and the sailing ship *Express*. These two ships carried a cargo of nails, tobacco, salt, petroleum and a metal lighter in parts; they discharged a short distance upstream. They loaded a cargo of 600 tons, principally of grain, and returned across the Kara Sea. The paddle tug *Moskwa* of Bremen under Captain Dallmann also made the voyage, and went right up the river to Yeniseysk, about 1400 miles upstream, and apparently remained on the river. Another small ship, the *Tsaritsa*, came down the Yenisey, went aground off the mouth of the river and was escorted across the Kara Sea by the *Fraser* and the *Express* (27). The *Moskwa* and the *Tsaritsa* belonged to Baron L. Knop, a Russian business man who had decided to follow the example of Sidorov and Sibiryakov.

Meanwhile Wiggins and others were making equally successful voyages to the Ob'. Wiggins took a cargo of manufactured goods and salt in the *Warkworth* of 650 tons burden. He did not cross the bar of the Ob' but was able to exchange his cargo for grain which had been brought downstream by river craft to Nadymskaya Guba, a point about 100 miles out in the estuary. River craft had been working on the Ob' system for a long time, probably longer than on the Yenisey. The *Neptun* of Hamburg made a similar voyage to that of the *Warkworth*, and carried a similar cargo. The *Sibir'*, a ship built by a Russian merchant at Tyumen' on the Ob' system came down the Irtysh and the Ob' and sailed through to London. Thus five ships came to the two rivers in 1878, and six sailed out of them to the west. The amount of freight carried was much greater than had ever before been the case.

The successful voyages of 1877 and 1878 encouraged speculators in England to hope that there was easy money in the Kara Sea shipping business. In

1879 the Russian Government was persuaded to grant a limited concession for the import of certain goods duty-free by way of the Ob' and Yenisey (81). Five large steamers were chartered for the voyage to the Ob'; 5000 tons of cargo was sent down to the mouth of the river to meet them. But none of the sea-going ships got further than Karskiye Vorota. The most elementary mistakes had been made: the ships were unsuited for navigation in ice, information about shoals and currents was ignored. Wiggins for these reasons had refused to take command of the convoy. The failure of this venture not unnaturally did much damage to the cause of popularising the route. But there is no doubt that ice was bad this year. The *Neptun* tried but failed to break through the ice east of Matochkin Shar. Captain Dallmann alone, in the *Luise*, was able to make the voyage to the Yenisey.

Nor was 1880 a much more successful year. Sibiryakov himself decided to accompany one of his steamers, the *Oscar Dickson*. He sailed with another ship freighted by him, the *Nordland*. His two ships went aground near the mouth of the Yenisey and were compelled to winter there. They subsequently became total losses (300). Knop again equipped a Yenisey expedition in this year. His ships *Luise* and *Dallmann*, commanded by Captain Dallmann, reached the Kara Sea. It is not clear what happened to them,[1] but neither seems to have reached its destination. The only ship to make the passage successfully was the *Neptun* which took a cargo to the mouth of the Ob' and returned with grain.

There is little information available on shipping in the Kara Sea in 1881. Nansen (204) lists two steamers and some lighters as having reached the Yenisey. It is unlikely that more vessels sailed. The succeeding years saw very little activity, evidently partly at least because ice conditions were bad. Two ships heading for the rivers failed to get through, and two expedition ships were caught in the ice and drifted in the Kara Sea all winter. Between 1882 and 1886 there is no record of any ship reaching the Ob' or the Yenisey.

During these years Wiggins, finding no financial support for voyages to Siberia as a result of the failure of 1879, was sailing in southern waters. In 1887 however a company was formed in England—the Phoenix Company—with the object of trading to the Ob' and Yenisey. For the next seven years this company and its successor had no competitors in the Kara Sea trade. Sibiryakov and Knop gave up after 1882. Through the good offices of the British Ambassador to Russia, the Phoenix Company obtained a concession from the Russian Government that goods might be brought in free of duty by this route. Wiggins was appointed the company's marine superintendent and took command of the steamer *Phoenix* of 273 registered tons. In 1887 the *Phoenix* made her way up the Yenisey to the town of Yeniseysk, unloading cargo at villages *en route*. She wintered at Yeniseysk, and it was intended that

---

[1] Sibiryakov (299) writes that the *Luise* turned back after being unable to find a way through the ice off Novaya Zemlya. Wiggins's biographer (90) mentions that the *Luise* was lost in Obskaya Guba together with two schooners. Yet a *Luise*, evidently the same ship, was trying to make the same voyage two years later (203, 262).

she should henceforth work on the river while the company sent other sea-going craft as far as the mouth.

In accordance with this plan the *Labrador*, a steamer of 391 tons gross, was loaded with mining machinery and set out the following summer for the Yenisey under Wiggins's command—he had returned overland to England. A concession for importing goods free of duty for five years on the Yenisey and one year on the Ob'—the biggest yet made—was secured, thanks to further action by the British Ambassador. Unfortunately the *Phoenix* went aground in the Yenisey while coming downstream to her rendezvous with the *Labrador*. A ship was sent from England to replace the *Phoenix*, but she was unable to weather the ice on her journey out and had to return. There was no object in the *Labrador* continuing the voyage since there was no vessel to carry her freight up the Yenisey (the *Labrador* herself drew too much water to make the voyage upstream). Wiggins therefore turned back. Ironically, the *Phoenix* was able to continue the voyage downstream after all, and reached Gol'chikha at the mouth of the Yenisey; but there was no way of letting Wiggins know this in time.

This disappointment was too much for the Phoenix Company. The concern sold out to the Anglo-Siberian Trading Syndicate, which planned a second trip for the *Labrador* in 1889. Wiggins was again in command. The Yenisey was successfully reached, but owing to a misunderstanding the river steamer which was to meet the *Labrador* waited for her at Karaul, while the *Labrador* was waiting 200 miles farther downstream at Gol'chikha. Once again therefore the expedition was an almost complete financial loss.

The syndicate determined to continue, and in 1890 it sent two shallow-draught cargo vessels, the *Biscaya* and the *Thule*, and a tug, the *Bard*, to the Yenisey. This time at last the sea-going vessels met the river steamers and exchanged cargoes. The *Bard* went up the Yenisey, while the other two ships returned to England. This trip was thoroughly successful both from the organisational and commercial points of view. It was a pity that Wiggins, who had done so much of the pioneer work, was not with the convoy. Several of his crew from the *Labrador*, however, had responsible positions in the ships that made this voyage.

During the next two years there was, surprisingly, no navigation at all in the Kara Sea. But in 1893 the Russian Government at last started to take an interest on its own account in the possibilities of the route. During the twenty years since Wiggins had first shown that navigation was possible in the Kara Sea, the Russian Government had done nothing, in spite of Wiggins's frequent attempts at persuasion, to encourage trade by that route, beyond allowing occasional exemption from import duty for limited periods. It had evidently never struck the Government that the route might one day become useful to the country as a whole. The construction of the Trans-Siberian railway was the factor which made clear the importance of the Kara Sea route. In 1891 work began on a line that was to run eastwards from the railhead at Zlatoust in Ural. The Yenisey was seen to be a quick and cheap route by which rails could be brought to Krasnoyarsk, a point on the upper Yenisey through which

PLATE I

CAPTAIN JOSEPH WIGGINS (1832–1905), one of the
pioneers of the Kara Sea route.

PLATE II

ALEKSANDR MIKHAYLOVICH SIBIRYAKOV (1849–1933),
a Russian merchant who was actively interested in promoting
voyages to the Ob' and Yenisey.

the railway was to pass. Accordingly, the Committee of the Siberian Railway approached Wiggins, who was planning a voyage in 1893. The voyage Wiggins was planning was to be undertaken in company with and on behalf of an English business man, F. W. Leyborne-Popham. The object of the trip was partly pleasure, partly a rather vague hope of trade, and partly in order to bring stores to Nansen who was about to start his voyage in the *Fram* along the north Siberian coast. Two ships were to be taken: the yacht *Blencathra*, and the *Minusinsk*, a shallow-draught vessel for river work on the Yenisey. The Committee of the Siberian Railway wanted Wiggins to escort to the Yenisey the steamer *Orestes* carrying 1600 tons of rails, the paddle steamer *Leytenant Malygin*, the steamer *Leytenant Ovtsyn* and the barge *Leytenant Skuratov*. The last three were all built in England for the Yenisey river fleet, and sailed under the command of Lieutenant (later Admiral) L. F. Dobrotvor-skiy of the Imperial Russian Navy. The journey was successfully completed. The rails were unloaded at Gol'chikha and the small craft went up the river. Nansen was found to have crossed the Kara Sea some days before the convoy arrived. The only misfortune was that there were too few barges at Gol'chikha to take all the rails, some of which had to be taken back to Arkhangel'sk and unloaded there. Wiggins went up the river and received the thanks of the Russian Government for the safe arrival of its goods and ships.

The Committee of the Siberian Railway again asked Wiggins's help in the following year, 1894. The Committee had had two paddle steamers, *Pervyy* and *Vtoroy*, built in England, and it wanted these brought to the Yenisey. Wiggins himself sailed in the *Stjernen*, a steamer acquired by Leyborne-Popham, who was now very interested in the prospects of trade. The convoy reached the Yenisey safely. On the return journey however the *Stjernen* was wrecked on an uncharted reef near Yugorskiy Shar. This was the first loss at sea that Wiggins had suffered in nine seasons in these waters. All hands were saved.

Leyborne-Popham was not discouraged by the wreck. In 1895 he bought the *Lorna Doone*, a barque of 381 tons gross, and had her converted to steam and strengthened for Arctic work. A cargo of manufactured goods was loaded. The *Burnoul*, a shallow-draught ship, was to accompany the *Lorna Doone*. Wiggins was again in command, and he brought the two ships to Gol'chikha successfully. The *Lorna Doone* loaded a cargo of flour and graphite from Kureyka, and returned. The *Burnoul* went up river, as had been planned, and wintered there.

Leyborne-Popham considerably enlarged his operations the next year. Wiggins was to take six ships: the *Lorna Doone*, the paddle steamers *Glenmore* and *Scotia*, the *Dolphin*, the *Mula* and the *Ioann Kronshtadskiy*. The *Glenmore* and the *Ioann Kronshtadskiy* had been bought by Siberian merchants for river work. There was divided responsibility for the convoy. Some of the ships were sent on ahead of Wiggins by their owners; consequently when Wiggins reached Vardø in northern Norway, the point of assembly, he found he was supposed to enter the Kara Sea without having with him the *Lorna Doone*, which was the only ice-strengthened vessel of the six. He refused to do this, and returned with four ships. *Lorna Doone* and *Dolphin* reached Gol'chikha. There was much ill feeling about all this, since it meant considerable financial loss instead of a large gain. Wiggins would appear to have been justified in his action, since he was responsible for the convoy and could not therefore brook any interference by people lacking his expert knowledge of ice navigation. The result of it all was that Wiggins severed his connection with Leyborne-Popham's syndicate.

During the next three years the syndicate continued to send out ships. In 1897 eleven ships steamed to the Ob' and Yenisey with freight. Grain was loaded for the return voyage, which was successfully accomplished. In 1898 four ships took 3000 tons of tea and some machinery to the Ob', and one ship and a schooner went to the Yenisey. In 1899 four large ships were sent out. Ice conditions were bad. One ship was damaged by ice near Yugorskiy Shar and sank, and the other three turned back. Although this failure could not, one would have thought, wholly offset the previous successes, it in fact put a stop to the operations of the syndicate.

After the failure of 1899, there were no more attempts to reach the Ob' or Yenisey for five years. The risks were thought too great by potential financial backers. Then in 1905 it was once again the Russian Government which saw a use for the route. The problem before the Government was how to relieve famine in central Siberia. The Russo-Japanese war had broken out, and the Trans-Siberian railway, completed in 1905, was loaded to capacity with war supplies for the Far East. It was decided to send a large convoy of food ships through the Kara Sea. Wiggins, whose reputation in Russia was high, was asked to take charge; 22 ships and lighters were to take part. There were last-minute difficulties, and in the end seven steamers and tugs and nine lighters successfully made the voyage, carrying 12,000 tons of cargo. The tugs and lighters remained on the river. The Russian Government's first large ice-breaker, the *Yermak*, completed six years earlier in England, was to accompany the convoy; but she went aground off Vaygach. Two German steamers went to the Ob' in the same year. Wiggins, to his own great disappointment, was prevented by serious illness from accompanying the convoy—which was, of course, the biggest that had yet sailed those waters. While it was still at sea, in September 1905, he died. His death, coinciding with the successful voyage of a large convoy through the Kara Sea, can be said to mark the end of the pioneering stage of the development of the Kara Sea route. It was Wiggins's many voyages which did more than anything else to dispose of the ice bogey. Merchandise totalling 20,000 tons, including much gold-mining equipment, is said to have been imported into Siberia by this route up to (but excluding) 1905 (92). This is certainly a conservative estimate.

Yet after the large expedition of 1905, there was another period of inactivity. With the Russo-Japanese war ended, the Trans-Siberian railway became available again for normal goods carriage into Siberia. The Russian Government evidently felt that the railway would be sufficient to keep Siberia supplied. Nothing therefore was done to encourage use of the Kara Sea route. Exemption from duty had been allowed to the incoming cargoes in 1905. This was only allowed on certain classes of goods in 1906, and was not granted at all in the years that followed. As a result, British business men were not inclined to risk capital any more than they had been after the failure of 1899. Russian merchants had never, with a few exceptions, been particularly keen on the idea. Sibiryakov, the chief exception, concluded well before the turn of the century that "shipping in the Kara Sea has great difficulties to contend with and therefore is not suitable for commercial use" (298). He turned his attention instead to inland waterways across Siberia, an idea he had always favoured. As a result of the general lack of interest, the only traffic to the Ob' or Yenisey between 1906 and 1910 was the voyage of one Russian steamer, the *Bakan*, to the Yenisey in 1907.

The year 1910 saw the beginning of another period of activity—the last attempt by foreign business interests to make a financial success of the Kara Sea route. Jonas Lied, a Norwegian business man, read of Wiggins's voyages and became interested. He spent the summer of 1910 in Russia, and went down the Yenisey by river steamer. He was able to interest some British business men in his plans. One of these, Valentine Webster, was excited by the idea and decided to try to take through a cargo to the Yenisey himself in 1911. His ship *Nimrod* reached the Yenisey successfully, but Webster had apparently not notified the Russians of his intended arrival. The customs authorities therefore arrested him and took charge of his cargo when both he and it finally arrived at Krasnoyarsk. Webster was released shortly afterwards and heavily fined, but he was allowed to sell his cargo. He took no further part in Kara Sea operations. While this was happening, Lied was again in Siberia making the necessary arrangements with merchants for his projected voyage of the next year.

In 1912 Lied made his first attack on the Kara Sea route. He formed the Siberian Steamship Manufacturing and Trading Company in Christiania, and chartered the steamer *Tulla*. A return cargo was carefully selected in Siberia and taken down the Yenisey to Ostrov Diksona. But the *Tulla* could not force the ice in the Kara Sea, so the whole expedition failed. Lied however was able to persuade his company not only to have another try but to increase its capital before doing so. In 1913 he chartered the steamer *Correct* of 1036 tons gross. He obtained a contract from the Department of Railway Construction and Development to transport 1000 tons of cement for the Altay railway. He also persuaded Fridtjof Nansen to make the voyage as a guest on board the *Correct*. The voyage was successful. The cement was replaced by some 550 tons of flax, bristle, wool, timber and graphite on the return journey. Nansen's presence, and his published narrative of the voyage in *Through Siberia, the land of the future* (1914), attracted attention to the event. Lied's company was firmly established. Lied himself became a Russian subject for business reasons.

The next year quite a large convoy sailed to the Yenisey under Lied's direction. Two ships, the *Ragna* of 1747 tons gross and the *Skule*, 1150 tons gross, were chartered. Two steel barges sailed from England, carrying the components of two river steamers. Four river steamers bought by the Russian

Government sailed from Germany the day before war broke out. The convoy assembled at Tromsø and sailed to the Yenisey. One German river steamer escorted one of the barges to the Ob'. By arrangement with the Russian Government an aircraft was to have been made available for ice reconnaissance—the first use of aircraft for this purpose. But the outbreak of war forced the cancellation of this plan. The commercial side of the expedition was very successful. All the previous losses of the company were more than covered. The Russian Government had again been the company's principal client. Besides the river steamers, Lied transported 5000 tons of cement for the railway, machinery and coal. The outgoing cargo consisted of timber, some thirty tons each of butter and Kureyka graphite, and the usual flax, hemp, hides and bristles.

The year 1915 was also successful. Lied chartered two Norwegian ships, the *Haugastoel* of 2118 tons gross and the *Eden* of 1304 tons gross. The *Eden* went to the Yenisey and the *Haugastoel* to the Ob'. Currency difficulties prevented the import of any cargo into Siberia; but the experiment of exporting butter the previous year encouraged Lied to take 30,000 casks of butter to England. In spite of the lack of refrigeration, the butter arrived in good condition. The capital of the company was again increased after this.

Lied had been active all this time at both ends of the sea route—in London and Christiania and also in Krasnoyarsk and Novonikolayevsk (now Novosibirsk). He wanted to continue expanding his business. In 1916 therefore he acquired control of one of the two river fleets on the Yenisey—the other was owned by the Government—and he acquired joint control with the Russian merchant Kornilov of the Ob' river fleet. Control of the river fleets was obviously an enormous advantage to the organiser of sea shipping which was entirely dependent on river craft for carrying cargoes to and from the market. The Yenisey fleet comprised ten powered vessels and thirty barges, and the Ob' fleet 49 powered vessels and 140 barges. Lied had further ambitious plans to develop his own sources of raw materials in Siberia. He bought a site for a shipyard in Krasnoyarsk, a canning factory on the lower Yenisey and a large area of standing timber at Maklakova on the Yenisey just above Yeniseysk. His company bought vessels of its own this year. One of these, the *Edam* of 2381 tons gross, was sent to the Yenisey. A Government ban on exports from Siberia was lifted for this ship, but it would have been impossible to get permission to send more than one ship. The *Edam* reached the Yenisey safely. One of the main items of the cargo was complete equipment for a sawmill to be erected at Maklakova. Flax was the cargo on the return journey, but the *Edam* was torpedoed off Bergen. In spite of this (for the ship was insured) the company made a 40 % profit on the year's trading. The company now had offices in St Petersburg, Moscow, Arkhangel'sk, Krasnoyarsk, Novonikolayevsk, Christiania, London and New York. The capital was doubled in the autumn of 1916.

In 1917 Lied planned to send the *Obj* of 1828 tons gross, owned also by his company, direct from the United States to the Yenisey. He wanted to persuade Theodore Roosevelt to accompany the expedition and hoped thus to achieve the same effect as he had with Nansen's voyage on the *Correct* in 1913. But the United States' entry into the war in 1917 put an end to the plan to get Roosevelt, and the strikes and confused political situation between revolutions in Russia in fact prevented the *Obj* from reaching her destination. She unloaded at Arkhangel'sk, loaded some potash for the return voyage and was torpedoed off the Murman coast. This was the last voyage undertaken by Lied's company. In 1914, 1915 and 1916 Lied, with his energy and business acumen, had been able to make a financial success of the Kara Sea route. Future prospects seemed very bright for him, but the Revolution prevented him from continuing his business and in fact ruined him. The proof which he had provided that the route could be successfully and profitably worked was by no means wasted on his successors, even if they refused to acknowledge their debt to him.

The Revolution caused no real break in the development of the Kara Sea route. Enough had been done by Wiggins and Lied to demonstrate its usefulness and practicability. The way was clear for large-scale development. What happened between 1917 and 1919 was that private enterprise was forced to yield its claim, allowing the state, if it liked the idea, to replace private enterprise.

Before we pass on to consider the post-Revolutionary period of state monopoly, it will be convenient to pause and examine the economic basis upon which private enterprise was able to function, and related facts. This is important if we are to get a clear picture of the objects and methods of operating the Kara Sea route during this period.

It is true to say that in the early years the emphasis was upon imports to Siberia rather than exports. The great need in Siberia was for manufactured goods. As far as Sidorov and Sibiryakov were personally concerned it was for gold-mining equipment. Both Sibiryakov and Wiggins's financial backers made the marketing of manufactured goods their chief concern. The low level of industrialisation in Russia and the remoteness of Siberia from manufacturing centres assured a ready market to such goods. Although imports to Siberia were the attraction to traders, return cargoes were always taken even if little attention was given to their composition or procurement. These cargoes normally consisted of grain, flax, bristles and hides, and occasionally small consignments of graphite(221). The agricultural produce came from central Siberia and even Turkestan. Generally such cargoes would be profitable; for instance, in 1877 a ton of wheat in the Tyumen' area is reported to have cost the same as a hundredweight in England(275). But if it had been a bad year in Siberia prices there would be high and the commodities would not be worth exporting. Sidorov, it is true, wanted to arrange the export of minerals— graphite and coal—from the Yenisey on a bigger scale, and spoke of exporting timber(354). But he was not able to do this.

Lied on the other hand evidently felt that one of the mistakes of his predecessors was to underrate the importance of the export side of the trade, and so he chose the goods for export with great care. He experimented with furs and very successfully with butter, and started exporting timber. The cost in

1909 of carrying one pud (36 lb.) of grain from Novonikolayevsk to London via Arkhangel'sk was estimated at 55 kopecks, via Petrograd at 67 kopecks, and via the Kara Sea at 28–35 kopecks(220). If the right commodities were chosen, there was a sound basis for business.

But there was another factor of great importance: the attitude of the state towards the project. The state is in a position to make or mar a new venture of this sort. It may either prohibit trading, as did Tsar Mikhail Fedorovich, or it may encourage it by giving assistance in varying degrees. The attitude of the Russian Government towards the Kara Sea route changed gradually from hostility to encouragement. In the middle of the nineteenth century the Government seems to have felt that the north was so remote and dangerous and so completely incapable of development that the correct course was to dissuade people from going there(353). Hence the difficulty experienced by Sidorov, Sibiryakov and others in implementing their plans. But the Government was indifferent if foreigners chose to risk their lives in the frozen north. This view however was soon made untenable by the facts. After some successful voyages had been made, the Government began to realise that the route might possibly be an asset to the country. Some concessions were granted to Kara Sea traders and some services were provided for them. The concessions took the form of remission of import duty on cargoes reaching Russia by way of the Ob' or Yenisey. But there was no firm Government policy in this. The tax was waived for short periods when special representations were made—as in 1879, 1887 and 1905—and was then reimposed. Reimposition was urged on the Government by an influential group of Russian industrialists who resented competition in the Siberian market(93). Once at least, in 1906, reimposition of the tax was probably a principal factor in halting traffic altogether. The concessions therefore had no permanent effect. In the provision of services however very real help was given.

The Government was moved to take active measures after the first voyage in which a Government department had participated. This was in 1893, when the Committee of the Siberian Railway had rails brought to the Yenisey. The urgent need for hydrographic information was realised. Hitherto the only chart of the region was one published in 1847 on the basis of eighteenth-century descriptions of the coast and the work of two hydrographers who investigated the coast from the Pechora to Yamal in the 1820's; there was later added the work of Moiseyev, another hydrographer, in Obskaya Guba in 1881(276).[1] Clearly no individual or even syndicate could undertake the expense of sending out the regular hydrographic expeditions which were necessary; this was a Government responsibility. In 1894 the Finance Minister, Count S. Yu. Witte, is said(91) to have written to the Emperor: "We must prove the route to be either a great Yes or a great No." The same year saw the establishment of the Hydrographic Expedition for the study of the mouths of the rivers Ob' and Yenisey and of part of the Kara Sea [Gidrograficheskaya ekspeditsiya dlya izucheniya ust'yev rek Obi i Yeniseya i chasti Karskogo morya]. This was not properly an expedition but an organisation in St

---

[1] According to Yu. M. Shokal'skiy(293) Moiseyev worked in 1888.

Petersburg which sent out expeditions. Its leader was A. I. Vil'kitskiy. At least one party was sent out each year for the next five years. The areas surveyed covered the estuaries of the Ob' and Yenisey, including a proposed trans-shipment point at Bukhta Nakhodka in the Ob' estuary; the Kara Sea littoral, chiefly in the region of Ostrov Diksona and Ostrov Belyy; the Pechora estuary; and parts of the Murman coast(295). The ships principally used were the *Leytenant Ovtsyn* and the *Leytenant Skuratov* which had been brought from England in 1893(294). In 1898 the name of the Hydrographic Expedition was changed to the Hydrographic Expedition of the Arctic Ocean [Gidrograficheskaya Ekspeditsiya Ledovitago Okeana]. The ship *Pakhtusov* was put at its disposal. Voyages were made each year until 1904, with the exception of 1903 when ice prevented the *Pakhtusov* from reaching the Kara Sea(204). In 1902 A. Varnek succeeded Vil'kitskiy, and in 1903 F. Drizhenko succeeded Varnek(295). After the 1904 season the Expedition ceased activities for a period. Work had continued in the same main areas as before. On the basis of the information collected in these eleven years charts and sailing directions were compiled for the voyage from Yugorskiy Shar to the Ob' and the Yenisey, and an atlas of the Yenisey up to Yeniseysk was prepared(16).

It is worth while mentioning here that in 1897 the Government sponsored, admittedly rather unwillingly, the construction of the powerful icebreaker *Yermak* on the grounds that such a vessel, if it proved a success, would greatly help the Kara Sea trade(154). The *Yermak* was in fact an excellent ship, but owing to unreasonable official disappointment in its behaviour on its first Arctic trials, it was not used in those waters until 1934 (see p. 71).

After the first burst of activity there was a break in the performance of hydrographic work in the Kara Sea area. The Government had seen the economic usefulness of the route and the importance of charting it. The Russo-Japanese war was fought, with disastrous results for the Russians and especially for the Russian fleet. It was in fact the difficulty experienced during the war of deploying the fleet and concentrating it in the Pacific that focused official attention once more on northern waters. Had it been possible to send naval units along the Northern Sea Route to the Pacific, the disaster of Tsushima might not have occurred. It was thus strategic rather than economic considerations which led to the official decision to investigate more thoroughly the whole length of the route. The Hydrographic Expedition of the Arctic Ocean renewed its activity. An ambitious plan was put forward. Sixteen polar stations were to be set up at intervals along the route, and three sea-going exploration and research groups, each consisting of two ships, were to work for three years in these waters(371). But in the end a more modest plan was adopted. Two small icebreakers were built and manned by naval crews. These two ships were to start work on the western end of the route, in the Kara Sea, in 1909. But it so happened that there were goods to be taken from European Russia to the Lena and Kolyma in that year, and it was decided to send the expedition and its two ships *Taymyr* and *Vaygach* to the Pacific in order to take the goods to their destination and start working from the eastern end(324). We will consider this important expedition later. The Kara Sea meanwhile was neglected. It was not until 1913 that anything further of importance was done. In this year a party in the Government hydrographic ship *Nikolay II* built navigation marks on Ostrov Belyy and Ostrov Vil'kit-skogo, and a permanent telegraph and meteorological station was set up at Yugorskiy Shar. In 1914 two similar stations were established at Mare Sale, on the west coast of Yamal, and on Ostrov Vaygach. In 1916 another was built at Ostrov Diksona. The primary duty of these stations was to report ice conditions during the navigation season, and in this they were very useful. It had long been clear that the establishment of such stations was the next step to be taken in making the route safe for navigation. More stations were planned(68)—for instance at Matochkin Shar and at the northern point of Novaya Zemlya—but the Revolution in 1917 caused a temporary delay. Meanwhile the continuous meteorological record kept by the four stations began to provide a foundation on which climatic theories could be built.

Thus before the Revolution a very definite start was made on measures to secure the safety of shipping in the Barents and Kara Seas. The sea lanes to the Ob' and Yenisey were already well enough charted for a wreck to be a rarity.

A striking feature of the pre-Revolutionary period is the leading part played in the development of the Kara Sea route by foreigners. The attitude of the Russian Government undoubtedly had much to do with this. The Government's disapproval of the northern aspirations of its own people in the early stages was followed by a period during which foreign investment in Russia was officially encouraged as a means of promoting industrialisation. Both I. A. Vyshnegradskiy and Count S. Yu. Witte followed this policy during their terms of office as Finance Minister (1887–1903). But it is also clear that Russian merchants after Sidorov and Sibiryakov showed a great lack of initiative in the matter. There was a certain feeling in the country against the foreign exploiters of Russian natural resources. At first opposition was small. But when it became clear that large profits were being made resistance seems to have stiffened. A keen Russian advocate of northern development wrote in 1914: "It is better, if there are no Russian enthusiasts, that these [Russian natural resources] should for the moment lie waste, rather than be plundered by foreigners"(212). But feeling of this sort was never a serious obstacle to Lied. More effective than this opposition by nationalists was the opposition by industrialists who resented competition and who, as we have seen, had influence with the Government. But before the issue really became serious the Revolution put an end to these opposing factions by upholding the nationalist case and crushing the industrialists.

We may insert here a word on the geography of the routes hitherto used by shipping in the Kara Sea. The termini in Siberia were the rivers Ob' and Yenisey, rather more frequently the latter. The route to both rivers lay through one of the three straits leading from the Barents Sea to the Kara Sea, and thence as directly as ice would allow to either Ob' or Yenisey estuary—Obskaya Guba or Yeniseyskiy Zaliv, as they are called. Serious thought was given to the plan of re-establishing the seventeenth-century route to the Ob' across Yamal.

Construction of a canal, which was necessary in order to avoid a portage, was planned to start in 1917 (78). But nothing came of this project. The Ob' has an 8-ft. bar, so that all but the smallest sea-going vessels had to trans-ship their cargoes in the estuary. This also is shallow. The points normally used for trans-shipment were first Nadymskaya Guba, on the southern shore of the estuary at the point where it turns north, and later Bukhta Nakhodka on the northern shore opposite Nadymskaya Guba. Both these places were about 200 miles from Obdorsk (now Salekhard), the most northerly settlement on the river itself, and there was little shelter from wind and sea at either place. The upper reaches of the Ob' were at the end of the nineteenth century more thickly populated than the upper Yenisey, since the Ob' was closer to European Russia. There was potentially therefore more trade to be done on the Ob'. But the Yenisey had the great advantage of a 24-ft. bar. Ships were able to go upstream to any one of a number of sheltered bays in order to trans-ship their cargoes on to river craft. Most of these trans-shipment points lay between Gol'chikha and Dudinka, both of which were themselves used. Some small sea-going ships went up the river as far as Yeniseysk.

Let us now take up again the main narrative where we left it, at Lied's last voyage in north Siberian waters in 1917. There were several attempts at using the Kara Sea route during the confused period of revolution and civil war between 1917 and 1920. The route acquired potential importance in 1918, when there were anti-communist forces in the region of Arkhangel'sk and also in Siberia. Arkhangel'sk lacked food, which was obtainable in Siberia, and the Siberian anti-communists lacked arms, which were available in Arkhangel'sk. The most direct means of communication between these two groups was by way of the Kara Sea. B. A. Vil'kitskiy, who had had a great deal of experience in north Siberian waters, was on the anti-communist side. He suggested using the route for an exchange of goods. He led an expedition in the *Taymyr* and the *Vaygach*, the ships in which he had traversed the whole Northern Sea Route in 1914–15. The object was to set up wireless stations at the mouth of the Yenisey. The *Vaygach* was wrecked at the mouth of the river, and the expedition was only a partial success. In the same year, 1918, the *Solombala* sailed from Arkhangel'sk to the Ob' carrying a mission which was to arrange the transport of war materials to Omsk. At Omsk also there was awareness of the importance of the route. The Ministry of Trade and Industry of Admiral Kolchak's Government formed the Directorate of Lighthouses and Sailing Directions of the Northern Sea Route, which became in April 1919 the Committee of the Northern Sea Route [Komitet Severnogo Morskogo Puti]. Lied was at this time in Siberia (he was still Russian by nationality) and he was trying to persuade Kolchak's Government to let him arrange for the import of British goods via the Kara Sea. But although Kolchak was willing, the British were not, and the plan fell through. Lied went over to the Soviet side soon afterwards.

Meanwhile at Arkhangel'sk in 1919 an expedition to the Ob' was fitted out. Nine ships, including an icebreaker, met river craft at the mouth of the Ob'. Although most of the cargo was successfully transferred to the river ships, only about a third of the return cargo was loaded because communist victories on the upper Ob' caused the trans-shipment to be completed in a hurry. The nine ships returned safely to Arkhangel'sk with some food from Siberia.

In 1920 all resistance to the communists was finally crushed. In April of that year the Siberian Revolutionary Committee formed another Committee of the Northern Sea Route [Komitet Severnogo Morskogo Puti, abbreviated to Komseveroput' or Komseverput']. The objects and scope of this Committee were broader than those of its predecessor. Its task was "all-round equipment, completion and study of the Northern Sea Route with the object of turning it into an artery of constant practical communication, and also technical organisation and development of an exchange of goods with overseas countries, and the transport of goods from European Russia by this route via the mouths of the Ob', Yenisey, Lena and Kolyma"(165). Thus on paper the Committee was responsible for organising sea-going traffic to and from four rivers; but in practice the Ob' and Yenisey remained by far the most important.

Komseverput' set to work, and grew steadily in importance. It had a monopoly in Kara Sea transport. In the early stages it was handicapped by lack of vessels and experience. Those who knew most about the route had left Russia. Lied, although he was living in the country and was now a Soviet citizen, was a capitalist and therefore not to be employed. These difficulties were gradually overcome. B. A. Vil'kitskiy, although he had emigrated, was actually employed for a time by the Soviet Government to take charge of convoys to the Kara Sea. For three years Komseverput' remained under the control of the Siberian Revolutionary Committee. In 1923 it was taken over by the People's Commissariat for Trade of the U.S.S.R. [Narodnyy Komissariat Torgovli SSSR, abbreviated to Narkomtorg], an indication of the economic importance attached to Komseverput'. In 1928 it was reorganised into a stock company with the name North Siberian State Stock Company "Komseverput'" [Severo-Sibirskoye Gosudarstvennoye Aktsionernoye Obshchestvo "Komseverput'"]. The shares were held by Narkomtorg and the Siberian Regional Executive Committee. The object of creating a company of this sort was to simplify the organisation of an undertaking which was concerned with more than one government department. Komseverput' was now concerned with trade and industry in addition to transport. The industries for which it became responsible in its own area were mining, fish preserving, hunting and the timber industry. Expeditions were sent out in the Yenisey area to study mineral deposits, timber resources and fisheries. The company's headquarters were in Novosibirsk (formerly Novonikolayevsk) and its river bases were at Omsk and Krasnoyarsk. By 1931 it had representatives in Moscow, Berlin and London. In 1933, however, the company was dissolved for alleged inefficiency and laxity in financial affairs, and its responsibilities were assumed by a government department created to accelerate the development of northern sea transport. The formation of this department marks the start of a new epoch in the treatment of Arctic problems in the U.S.S.R., and this will be dealt with in Part II of this study.

It is no longer necessary to recount voyages to the Kara Sea year by year. The stage had been reached at which the element of danger no longer played an important part and the yearly voyages became increasingly a matter of routine.[1] The number of cargo ships employed varied considerably from year to year but by the period 1929–32 a very marked increase is apparent.

The great majority of sea-going ships which used the Kara Sea route were foreign merchantmen, principally British or Norwegian, on charter to the Soviet Government. Komseverput' had no freighters of its own. Any Soviet ships used came from the Soviet merchant fleet which was run by the People's Commissariat for Water Transport [Narodnyy Komissariat Vodnogo Transporta, abbreviated to Narkomvod].

A number of aids to navigation were provided during this period. The usefulness of the icebreaker had long been realised. During the war of 1914–18 the Allies transferred a number of icebreakers to the Russians in order to keep open the supply line to the northern ports. Many of these ships were thus available for use after the Revolution. They were quite frequently used to assist the Kara Sea convoys. The first occasion was in 1921 when the *Lenin* was used as an escort to the freighters. During the years that followed an icebreaker was used as often as one was available. From 1929 one was used every year (379). The commander of the sea operations was usually either aboard this icebreaker or at Ostrov Diksona. The use of icebreakers meant less delay and greater safety for the merchantmen. Other measures to counter difficulties caused by ice were taken. An important one was the use of aircraft. The first flight in the Russian Arctic had been made in 1914 with the object of searching for survivors of an expedition lost in 1912; but sea-ice reconnaissance from the air for the benefit of shipping was first undertaken in 1924, when a hydroplane based at Matochkin Shar gave useful information to the Kara Sea expedition of that year. In 1925 two such aircraft did similar work. In 1929 Komseverput' formed its own air section and acquired one flying boat. In 1930 this number grew to three, and in 1932 to five. The main bases were Ostrov Diksona and Bukhta Varneka. Ice reconnaissance was the principal, though not the only, task of these aircraft (398). During the period of Komseverput's activity hydrographic work was continued and more polar stations were set up. An account of this will be given later in the section dealing with scientific work (see pp. 32–36). Vize claims (379) that as a result of all these measures, no vessel between 1920 and 1931 had to turn back because of insuperable ice in the Kara Sea.

The river fleets remained a vital part of the system. There was already, as we have seen, quite a considerable fleet on both Ob' and Yenisey, but this was chiefly for use on the upper reaches and many of the craft were taken over after the Revolution by inland water transport organisations having nothing to do with Kara Sea traffic. Komseverput' set about building up its own fleet. Initially any craft that were available were used, and they were not always suitable for carrying the increasing volume of goods. For the Ob' Komseverput' ordered construction of some barges of special design. They were able to carry 3000 metric tons of freight on the river and 2340 metric tons in the estuary. Their draught was 2·5 metres when fully loaded. Two such barges

[1] For details of Kara Sea traffic from 1920 see Appendix I.

were launched in 1924; there were seven by 1930. The first powerful tug to be acquired by Komseverput' for the Ob' was the *Sibraykom VKP(b)* of 1500 h.p. This came into service in 1928. For the Yenisey Komseverput' obtained two German-built tugs of about the same power. Large barges were also built for the Yenisey; notably three timber-carrying barges carrying 2600 metric tons each were built for the 1929 season (388). These were only the new craft. It seems likely that at any time between 1921 and 1932 there was on each river a fleet capable of carrying up to 7000 tons. By 1933 certainly there were said to be four powered craft and ten barges, with a capacity of 15,000 tons, on the Ob'; and on the Yenisey, 48 powered craft and 56 barges with a capacity of 16,000 tons (151).

The sea routes used by Kara Sea traffic during the period were substantially the same as those used previously. All three straits into the Kara Sea were used regularly. In 1930 for the first time two returning freighters sailed round the northern tip of Novaya Zemlya. Up to 1929 all the ships bound for the Ob' or Yenisey normally sailed in one convoy; there were never in fact more than eight ships involved. From 1929, when 26 ships were employed, it became necessary to arrange several separate convoys. This made the organisation more difficult, since there was generally only one icebreaker available for escort work. There were changes in trans-shipment points on each river. On the Ob' the 8-ft. bar was still an insuperable obstacle to sea-going ships; but a rather more sheltered bay than those previously used was found in the estuary about 35 sea miles north of Bukhta Nakhodka, the bay used by Lied and others. The newly discovered anchorage, called Novyy Port, was used from 1921 onwards. It was not really very satisfactory. It lay about 140 sea miles from the mouth of the river. The water sheltered by the arms of the bay had a depth of only 10–12 ft., and was unprotected from the east. River craft could shelter there after their 140-mile sea voyage, but the trans-shipment of cargo had to be carried out about three miles offshore where 18 ft. of water could be found for the sea-going vessels. Time was often lost while both river and sea-going vessels waited for the wind to drop sufficiently to allow trans-shipment to begin. But it was difficult to see any alternative to this plan; and after all at Bukhta Nakhodka sea-going ships could not approach nearer than twenty miles. Quite substantial tonnages could be exported annually because the river craft were able to make two journeys down- and upstream in a summer. On the Yenisey deep water at the bar and in the fairway allows sea-going ships to go a considerable way upstream. In the early 1920's Ust'-Port (also known as Ust'Yeniseyskiy Port and Ust'-Yeniseysk) was the most frequently used trans-shipment point. But northerly winds caused very high tides, which made it difficult to establish a shore base with storage warehouses. In 1927 a ship stopped at Igarskoye, a group of wintering huts on the right bank of the river some 360 sea miles upstream from Gol'chikha. Igarskoye was protected from the main stream by an island. The channel between them was over 400 yd. wide, had a depth of between 17 and 56 ft. in the fairway and a current of about half a knot. Sea-going ships could without difficulty penetrate this far upstream. Igarskoye seemed to be the ideal trans-shipment point. It was first used as such in 1928. From 1929 all vessels coming to the Yenisey from the Kara Sea used the new port. It was decided to build here sufficient shore installations to make the place, now called Igarka, a well-equipped base for Kara Sea operations. Sawmills were quickly put up, and Igarka became the focal point of the new and rapidly growing Yenisey timber export industry. As Igarka grew, so the Yenisey eclipsed the Ob' in importance. In 1921 four out of five ships went to the Ob'; in 1927, 1928 and 1929 about half the total number of ships went to each river; in 1932, 25 out of 28 ships went to the Yenisey (224, 279).

The Kara Sea route continued in this post-Revolutionary period to owe its existence to economic demands. The goods carried at the start of the period remained substantially the same as they had been before. Grain, timber, flax, hemp, fibre and bristle were the principal exports; machinery, chemicals and partly finished manufactured goods were the principal imports (96, 105). Imports did not show the increase that the greater number of ships used each year would have allowed. This was partly because industrial centres were growing up east of Ural in the southern part of the Ob' basin, and the railway network in western Siberia was enlarged. It was expected however that growing Siberian industry would in time call for bigger imports, especially of such things as the Kola apatites, which would be most easily transportable by sea (389).

The most striking feature of the freight turnover was the growth of exports. It became clear that the Kara Sea route's real usefulness to the Soviet state lay in the export of goods which could not be moved economically in any other way. Experiments were made in exporting the most valuable product of the Soviet Arctic, fur. But it was found that this compact and easily transported commodity could without difficulty and with greater speed be taken overland or by air. The table below shows the prices (given in roubles and kopecks) in 1927 in London of various commodities from Siberia, carried via the Kara Sea and via Leningrad (386):

Commodity	Via Kara Sea	Via Leningrad
Grain, per metric ton	24·06	24·94
Fibre, per metric ton	74·55	82·75
Pine oil, per metric ton	105·75	137·44
Timber, per standard of 165 cu. ft.	69·56	151·04

It is quite clear from these figures that timber was much the most advantageous cargo. The supply of timber from the Yenisey basin was enormous. Timber took up no space on river craft, since logs could be rafted downstream to Igarka where they were sawn and loaded direct on to sea-going ships. Timber in fact was the ideal export, for it was bulky and therefore uneconomical as railway freight, while it could be transported without difficulty by river, which was the north's most satisfactory communications system. Very naturally therefore Komseverput' concentrated on timber, and soon it became by far the largest export. Lied had been the first to export timber in any quantity, and had set up a sawmill at Maklakova in order to be able to develop his idea. But timber was not exported again until 1924. Later the sawmills at Igarka came into use and in the year 1928–29 81,000 cubic metres of timber were rafted downstream to them. In 1932 this quantity had grown to 422,000 cubic metres (155). From 1928 exports exceeded imports regularly, and by a large margin. The growth of timber exports was highly significant, since it altered the character of the Kara Sea route. Up to now it had been a new freight route to and from the interior of the country—cheaper than other routes but liable to be eclipsed by them if Siberia were to be intensively developed; now it became a route to indigenous northern resources for which water transport was the most economical—a position virtually unassailable by any future development of communications in Siberia.

Besides timber, there were of course other exports. These were principally the same agricultural products as had been exported since Wiggins's time. The export of fish was revived on a larger scale. Fishing in the estuaries of both rivers was encouraged and the catch was available at the right time and place to be loaded an to the ships.

As the volume of goods carried increased and more and more ships sailed to and from the Yenisey without mishap, the cost of freightage steadily dropped. Initially the cost had been high, chiefly because insurance rates were high. But as confidence increased the insurance rates came down. In 1914 these were 8 % on the ship and 6 % on the cargo; in 1929 2·75 % and 0·8 % (385). Besides the fall in insurance rates, another factor which helped to lower freight charges was the decrease in the length of the charter. The average charter for the journey from a western European port to the Yenisey and back was 81 days in 1924, 108 days in 1926 (when there were delays and difficulties with ice), and 70 days in 1930. Ships going to the Ob' in 1930 were chartered only for 61 days (141). The time taken for the voyage from western European port to Ob' or Yenisey was said to average sixteen days over the period 1921–29 (389). It was the turn-round which wasted most time, so there was reason to hope that improved port facilities would reduce the charter still further.

It is relevant to consider again the question of import duty. The Soviet Government had initially granted to Komseverput' the right to import goods free of duty. This was done, in rectification of what was considered to be a mistake of the Tsarist Government, in order to encourage trading and to compensate importers for the high insurance rates. In 1925 however freight rates had fallen to such an extent that payment of import duty was again enforced (384). But, unlike an earlier occasion, this action had no visible effect on traffic, since the state also ran the latter.

Since the whole enterprise directed by Komseverput' was a state monopoly, it is difficult to establish to what extent the increase in traffic was due to the fact that the Kara Sea route was a paying proposition. It is impossible to consult Kompseverput's accounts because one of the reasons for the company's dissolution in 1933 was that no accounts were being kept and that it was therefore impossible for the Government to determine the effectiveness of the investments made (314). This does however make it clear that the Government did subsidise Komseverput', and probably on a considerable scale. On the other hand Komseverput' had to undertake work not directly related to the Kara Sea route—for instance, development of natural resources on land, and scientific and exploratory work—none of which it could be expected to do without financial assistance. It seems on the whole likely that the running of the sea route was an economic proposition, provided the debit side of the balance-sheet included only costs of services concerned directly with the working of the route. Lied's experience of big profits with only two or three ships certainly leads one to believe that this would be the case.[1]

2. THE ROUTE FROM THE PACIFIC TO THE KOLYMA AND LENA

At the Pacific Ocean end of the Northern Sea Route commercial navigation did not start until much later than was the case at the Atlantic end. Wiggins and Nordenskiöld had brought the first samples of manufactured goods to the Yenisey in 1876, but no merchandise reached the Kolyma until 1911. There were various reasons for this. While the Kara Sea was relatively close to both Russian and European markets and production centres, Bering Strait was very remote from them. There was no economic need at the end of the nineteenth century for a route between the eastern Siberian rivers and places on the north Pacific seaboard. Further, the distance from Bering Strait to the large rivers is greater than that from the entrance of the Kara Sea to the Ob' or Yenisey. Though the Kolyma, which is the first large river west of Bering Strait, is not much farther than is the Yenisey from Karskiye Vorota, the Kolyma flows past no towns, forests or farmlands comparable to those on the Yenisey. The only river of north-eastern Siberia that is at all comparable to the Ob' or Yenisey is the Lena, which enters the sea about 1400 miles from Bering Strait, or almost as far again beyond the Kolyma.

Both Russians and foreigners had occasionally made voyages to the Chukchi and East Siberian Seas since the cossack Semen Dezhnev first sailed from the Kolyma through what was to become Bering Strait to the Anadyr' in 1648–49.[2] Almost the whole north coast of Siberia as far east as the Kolyma was visited by detachments of the Great Northern Expedition between 1733 and 1742. East of the Kolyma, perhaps the most important expedition from the point of view of mapping the coastline was that of 1820–24, led by F. P. Vrangel' and P. F. Anzhu. The coast from the Kolyma to a point less than 200 miles from Bering Strait was surveyed from the land side by Vrangel', while Anzhu and his party worked principally in the Ostrova Novosibirskiye. From the 1860's until 1930 American whalers and fur traders made a number of voyages to these waters. The traders established a very profitable fur trade with the natives of Chukotka. But as might perhaps be expected no hydrographic data were recorded. The *Vega* sailed along the whole coast in 1878–79, and thus showed the world that it was possible to sail quite a large ship from the Lena or Kolyma to Bering Strait. Nordenskiöld concluded on the basis of his voyage in the

[1] The principal sources of a general nature on which this section on the Kara Sea route is based are as follows: early history, 369; Wiggins's voyages, 89; Nordenskiöld's voyages, 150; Lied's voyages, 156; events of 1918–20, 164; general works, 85, 148, 200, 384.

[2] According to F. A. Golder(70), Dezhnev made most of this journey by land. But L. S. Berg(20) shows that the more widely held view that Dezhnev did go all the way by sea is probably right.

Vega that in most years it would be possible for ships drawing about 12–14 ft. to sail from the Pacific to the Lena. He considered however that further exploration was necessary before any trade route could be established (213). But it was thirty years before the necessary exploration was undertaken.

River traffic had sailed on the Lena for some time before communication with the outside world was made by sea. In 1856 the first river steamer was built on the river. The fleet grew until in 1917 there were said to be 37 powered craft and 107 barges on the Lena system (111).[1] These craft plied principally between the head of navigation, which was the closest point to Irkutsk and the

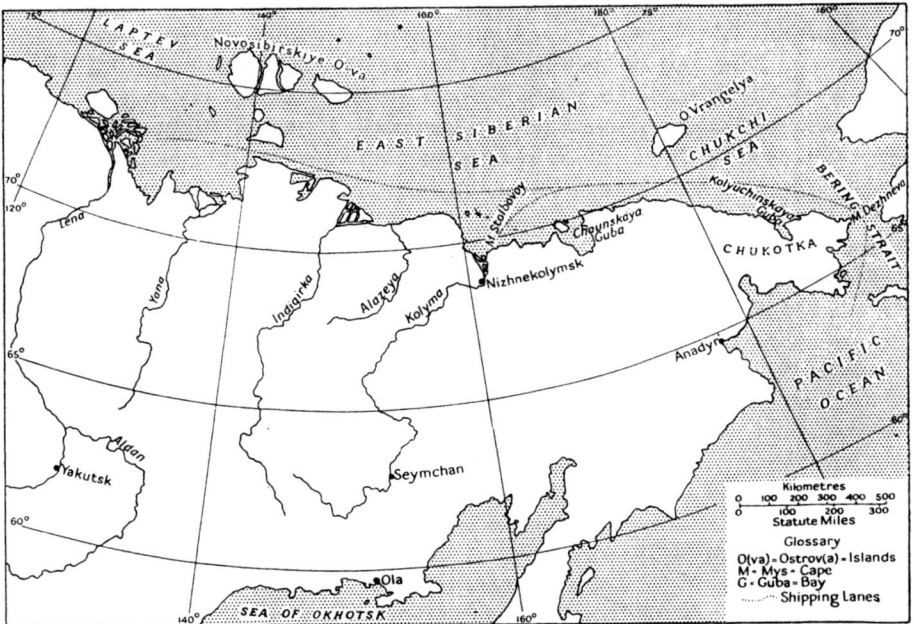

Map 2. The route to the Kolyma and Lena

main road (later also the railway), and the growing gold-mining settlements on the Vitim and the Aldan, two tributaries of the Lena. The upper reaches of the Lena provided the only highway—by water in summer, on ice in winter— between the whole of Yakutskaya Oblast' and the rest of Russia. Few of the river steamers were able to negotiate the wide and often stormy stretch below Yakutsk to the sea.

Since there was no obvious commercial advantage to be gained by pioneering a sea route for trade between the Pacific and north-eastern Siberia, in the end it was the Russian Government which started things going, and kept them going. The Government was concerned at the expense of transporting overland essential supplies to the settlements on the Kolyma and in Yakutskaya Oblast'

[1] A. Margolin (167), who mentions that there were only three powered boats in 1917, must presumably refer to the lower river.

in general. The route used lay through Irkutsk, down the Lena to Yakutsk, and thence by reindeer the remaining distance to any destination in the province—600 miles to the upper Kolyma, for instance. An alternative route to the Kolyma and Indigirka region lay through Vladivostok to Ola on the north shore of the Sea of Okhotsk, and thence overland some 250 miles to Seymchan on the upper Kolyma; but this route was found to be not less expensive and uncertain than the other (190). A commission was accordingly appointed in 1908 to examine the possibility of a sea route from Vladivostok to the Kolyma and Lena. The commission decided that some preliminary work was necessary before a shipping route could be established. In 1909 therefore three expeditions went to work in the area. I. P. Tolmachev led a topographical and geological expedition along the north coast of Chukotka from the Kolyma to Bering Strait; K. A. Vollosovich did similar work further west, between the Alazeya and the Yana; and G. Ya. Sedov surveyed the mouth of the Kolyma (211).

In 1910 the first large-scale hydrographic expedition ever to study north Siberian waters started work. This was the expedition on the *Taymyr* and the *Vaygach* which, as we mentioned earlier (see p. 15), unexpectedly started its study of the whole Northern Sea Route at the Pacific end. No doubt the Commission had been influential in arranging for the ships to come to this end. The expedition was led first by I. S. Sergeyev and later by B. A. Vil'kitskiy. From 1910 to 1914 the two ships made an expedition each summer from their base at Vladivostok, and during this period collected a great deal of information on offshore waters and islands in the Chukchi Sea, the East Siberian Sea and the Laptev Sea.

Meanwhile in 1911 the first voyage by a freighter to the Kolyma was made. The ship *Kolyma* sailed from Vladivostok with supplies from the inhabitants of the Kolyma valley. She unloaded at Mys Stolbovoy, outside the bar of the river, and returned the same summer to Vladivostok. Also in 1911 the American motor-schooner *Kittiwake* reached the river and sailed upstream to Nizhne-Kolymsk with her cargo. There was no organised river fleet on the Kolyma at this time, so the cargo was taken to its destination chiefly by reindeer. During the years that followed, some tugs and barges were brought from Vladivostok. In 1912 another Russian vessel, the *Kotik*, brought cargo to Nizhnekolymsk. Each year up to 1917 this voyage was successfully completed by either the *Kolyma* or the *Stavropol'* (which was the *Kotik* renamed). No voyage was made in 1918. In 1919 the *Stavropol'* and also an American schooner were prevented by ice from reaching the Kolyma. American schooners sailed in 1920, 1921 and 1922, but were not concerned in the plan for bringing supplies for the area. Between 1923 and 1930 American boats sailed quite frequently to the coast of Chukotka, but only occasionally did they reach the Kolyma. Each year from 1923 to 1927 the *Kolyma* or the *Stavropol'* sailed for the Kolyma; but in 1924 ice prevented delivery of the cargo. The amount of freight carried each year since 1911 only twice exceeded 500 metric tons and averaged considerably less.

In 1927, while the *Stavropol'* was going to the Kolyma, the ship *Kolyma* made a successful voyage to the Lena and back. Although a shipping route to the Lena had been envisaged ever since 1908, the *Kolyma* was in fact the first freighter to make the voyage. It was not repeated for several years. Each year from 1928 to 1932 the Kolyma service was continued and grew bigger; 2000 tons were carried in two ships to Nizhnekolymsk in 1931. In 1932 10,000 tons were carried in a large convoy of six freighters, four barges on tow, a schooner and two tugs for work on the Kolyma. Icebreaker escort was provided for the first time in the shape of the icebreaker *Litke* (399). Also for the first time in these waters aircraft were used for ice reconnaissance. Much of this last cargo however was not unloaded through bad organisation of the river side of the operation and bad weather. The river fleet, hitherto very small, acquired the barges and tugs. The goods carried on the voyages to the Kolyma from 1911 to 1932 were almost all imports to Siberia; at first essential supplies for the local inhabitants, later (in 1932) equipment for the gold mines then being developed on the upper Kolyma. The ships returned in ballast.

The voyages to the Kolyma and Lena were not undertaken without mishap. The conditions of sailing were comparable to those on the Kara Sea route thirty years earlier. After the *Taymyr* and *Vaygach* expedition, almost the only hydrographic work to be done concerned the rivers. Ships got no help from either icebreakers or aircraft until 1932. Although several flights were made in the vicinity between 1926 and 1929, their object was pioneering of air routes and not aid to shipping. There were no regularly working wireless or meteorological stations in the area until 1926; in 1931 there were only three stations on the 1400 mile stretch of coast between the Lena and Bering Strait. It is not surprising therefore that, as with the Kara Sea route, ships had difficulty with ice. We have already seen that ships were forced to turn back in 1919 and 1924 before reaching their destinations. In addition a number of ships were forced to winter at sea on the return voyage: the *Kolyma* in 1914, the *Stavropol'* and a schooner in 1929, two ships in 1931, and the whole convoy of six ships and escorting icebreaker in 1932. In 1931 the schooner *Chukotka* was crushed by ice *en route* to the Kolyma, and sank; this was the only occasion on which a ship was lost. Many of these accidents can be attributed to human rather than natural causes. In particular it may be mentioned that the *Kolyma* and the *Stavropol'* were both badly suited to Arctic work, a fact of which their captains frequently and unsuccessfully complained.

In the first twenty years of its working, the sea route to the Kolyma and Lena worked about as well, and claimed about as many casualties, as the Kara Sea route during its first twenty years. The results showed that it was possible to run a regular service, certainly as far as the Kolyma; but that there was plenty of room for improvement, particularly with regard to study of the behaviour of the sea, weather and ice.[1]

[1] This section is based on the following sources: 85, 318, 369.

3. SHIPPING IN THE WATERS BETWEEN THE YENISEY AND THE LENA

We have seen how ships started to ply between the Ob' and the Yenisey and the Atlantic in the last quarter of the nineteenth century; and between the Kolyma and Lena and the Pacific from 1911 onwards. Let us now consider the central part of the north Siberian coast, between the Yenisey and the Lena.

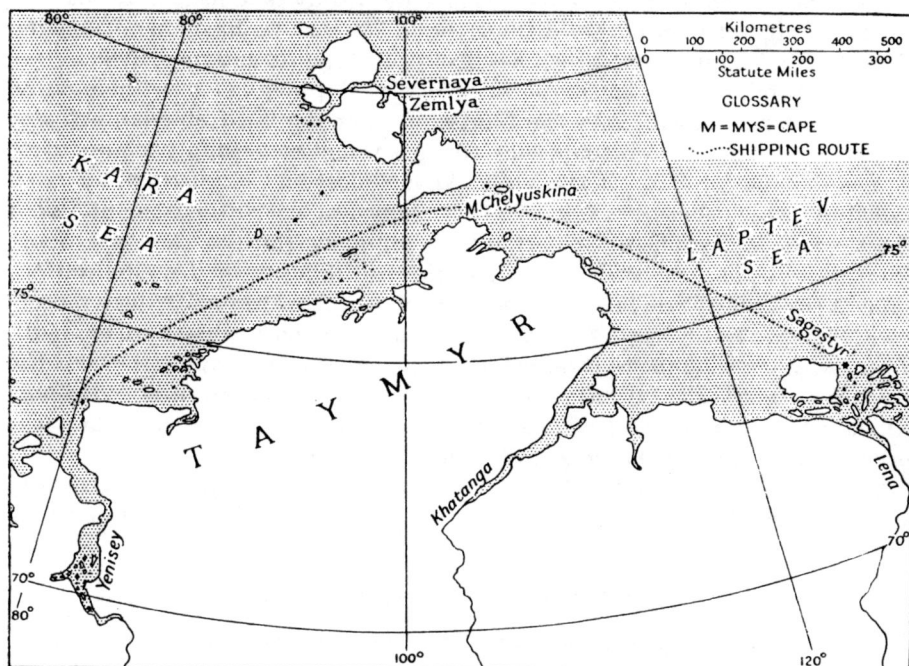

Map 3.   The central section of the Northern Sea Route

Almost the whole coastline had been described in general terms by members of the Great Northern Expedition between 1735 and 1741. Later a number of explorers sailed down the Yana or the Lena and visited Ostrova Novosibirskiye, but no further mapping of the coast or study of the rivers and seas between Lena and Yenisey was done until the end of the nineteenth century. In 1875 and 1893 geological parties under A. L. Chekanovskiy and Baron E. Toll worked on the coast between the Lena and the Khatanga, and between 1905 and 1913 the estuary of the Khatanga was again visited several times by I. P. Tolmachev and N. A. Begichev. A scientific station was set up at Sagastyr' in the Lena delta in 1882 and worked for two years. This station provided the first reliable data on weather and ice conditions in the Laptev Sea.

The first ship from the outside world to cross the Laptev Sea was Nordenskiöld's *Vega* in 1878. At Sibiryakov's request Nordenskiöld took with him a river steamer, the *Lena*, which parted company with the *Vega* off the Lena delta and steamed up the river to Yakutsk. The *Lena*, which belonged to Sibiryakov, was the first ship to reach the river from outside, and its arrival demonstrated the possibility of a sea link between the Lena and the Atlantic. But no one attempted to follow up this successful beginning. The pioneers of commercial shipping in north Siberian waters were fully occupied in the Kara Sea, and after all no one was likely to attempt the much longer voyage to the Lena at least until the route to the Yenisey was firmly established. In the end, the first freighter to reach the Lena came in 1927, as we have seen, from the Pacific. No freight reached the river from the Atlantic end until 1933.

The possibility of organising local coastwise shipping between the mouths of the big rivers flowing into the Laptev Sea and the East Siberian Sea was examined in 1923. The object of such voyages would be to supply remote parts of Yakutskaya A.S.S.R. (as the Yakutskaya Oblast' of Tsarist times was now called), in which the rivers were the best highways. Very little came of this in the next ten years. There were no suitable sea-going ships on the Lena, apart from the old *Lena*. Several expedition ships sailed out of the Lena to destinations in the Laptev Sea, but the only instance of freight transport was a motor-boat which towed a barge from the Lena to the Yana and back in 1932 (341). There was one attempt to establish contact between the Laptev Sea and the Kara Sea; this was the unsuccessful voyage in 1931 of the motor-boat *Belukha* which attempted to sail from Arkhangel'sk to the Lena and back but was prevented by ice from passing Mys Chelyuskina (108).

The only other ships to sail these waters were expedition ships which generally were not directly concerned with work in this area. We will consider the importance of these expeditions in the next section.[1]

4. THE NORTHERN SEA ROUTE AS A WHOLE

Having considered shipping in each of the three sections into which we have divided the waters north of Siberia, it remains to be seen how the idea of a sea route along the whole length of the north Russian coast arose and developed.

The first attempts to find a North East Passage were made by Englishmen in the middle of the sixteenth century. With the object of discovering a new route, free from Spanish interference, to the riches of Cathay, expeditions sailed under Sir Hugh Willoughby, Richard Chancellor and Stephen Burrough between 1553 and 1556; under James Bassendine in 1568 and Arthur Pet in 1580. Between 1594 and 1597 Dutch expeditions, on which Willem Barents was the leading personality, also attempted to find a North East Passage. From 1607 to 1611 Henry Hudson made a series of voyages in search of both North East and North West Passages. Although each one of these expeditions was remarkable, and some very sensational discoveries were made, none of the voyages came anywhere near to achieving its object; in fact none penetrated farther east than the western part of the Kara Sea.

After the early lack of success in finding the north-eastern route to Cathay, there were only sporadic attempts to do so during the next three centuries.

[1] This section is based on the following sources: 98, 215, 369.

Two English vessels under Captain Wood and Captain Flawes tried in 1676, but these did not pass Novaya Zemlya. After nearly a century a Russian expedition was sent out in 1765, at the suggestion of the scientist Mikhail Lomonosov, to see if a route to the Pacific could be found by way of the northern seas. This expedition, under V. Ya. Chichagov, made two attempts but did not get farther than Spitsbergen. Another Russian party under Fedor Rozmyslov set out in 1768 with the heavy programme of making a thorough study of Novaya Zemlya, sailing thence to the Ob', and then continuing to North America along the coast. For various reasons Rozmyslov got no further than Novaya Zemlya, where he did excellent work. In 1818 two English ships under Commander David Buchan, R.N., and Lieutenant John Franklin, R.N., attempted to reach Bering Strait by way of Spitsbergen, but ice forced them to return while they were still in the region of the latter. In 1872 the forcing of the North East Passage was one of several alternative objects of Weyprecht and Payer's Austrian Arctic expedition. The intention was to head east-north-east from the northern end of Novaya Zemlya with a deliberately vague plan to explore "unknown arctic regions". In fact the expedition's ship was frozen into the ice north of Novaya Zemlya and drifted north-westwards to unknown land which was named Kaiser Franz Josef Land.

The stumbling-block for almost all these expeditions had been ice conditions which were too severe for the ships used. In course of time ships were built which were better able to sail in ice-filled waters; and more was found out about the behaviour of sea ice generally. Thus in 1878–79, after the first exploratory voyages had been made to the Yenisey, Nordenskiöld in the *Vega* was able to succeed where all his predecessors had failed, and completed in two seasons a voyage along the whole length of the north Siberian coast. Besides having a good ship, Nordenskiöld had the two great advantages of his own skill and experience and good ice conditions. If he had not stopped at various places *en route* in order to do the scientific work which was one of his principal objects, he could almost certainly have reached the Pacific in the first summer of the voyage. On his return, however, he was cautious about prophesying a brilliant future for the Northern Sea Route. "Can the voyage of the *Vega* be repeated every year? It is for the moment impossible to answer this question either with a categorical yes or a categorical no....But I do believe that our voyage can be repeated often, and probably often with success" (214). As regards commercial navigation, he considered "it is not very likely, in the light of our present knowledge of the arctic seas of Siberia, that this route should *in its entirety* acquire effective importance for trade" (214). He thought that communication which could be established between the mouths of the great Siberian rivers and the seas of the world would have much greater importance than the Northern Sea Route as a whole.

Scientific expeditions sailed to these waters, however, and three should be mentioned. Each traversed a considerable section of the Northern Sea Route, and although none was specifically concerned with the opening up of a new sea route, the information obtained by all of them was relevant to that problem. In 1879, two months after Nordenskiöld sailed triumphantly through

Bering Strait into the Pacific, G. W. de Long, an American, sailed the *Jeanette* through Bering Strait in the opposite direction. The *Jeanette's* objective was the North Pole, since a theory was at this time current that the central part of the Arctic Ocean might be ice-free. She was frozen into the ice while still in the vicinity of Ostrov Vrangelya, and she drifted for nearly two years until she was finally crushed by ice and sank north-west of Ostrova Novosibirskiye in June 1881. Most of the crew reached land at the Lena delta after great exertions, but many, including de Long, died there. The *Jeanette* was the first ship to cross the northern part of the East Siberian Sea (36). Not long after the loss of the *Jeanette* a piece of wreckage which belonged unmistakably to her was washed up on the coast of Greenland. This was brought to the attention of Nansen, who determined to utilise the trans-polar current to which the piece of wreckage testified, and to drift with it right across the Arctic Ocean. In 1893 he set out in the *Fram*. He sailed from Norway along the north Siberian shore as far as Ostrova Novosibirskiye, where he turned north and allowed the *Fram* to be frozen into the sea ice. The *Fram* then started her three-years drift which followed the course Nansen had expected (199). In 1900 the *Zarya*, carrying a Russian Academy of Sciences expedition led by Baron E. Toll, sailed from the west to Ostrova Novosibirskiye, following the course of the *Fram* as far as these islands. The expedition worked there until 1902, when it returned, after losing Toll and three companions, by way of the Lena (340). Meteorological and ice information gathered by each of these voyages helped to build up a general picture of what a navigator might expect to find in the remoter parts of the Northern Sea Route.

There were two unsuccessful attempts in 1912 to repeat what the *Vega* had done. Both were simply voyages of exploration, and neither had commercial motives. Lieut. G. L. Brusilov in the *Sv. Anna* (which was Leyborne-Popham's *Blencathra*) set out from St Petersburg with the intention of reaching the Pacific Ocean by way of the Kara Sea. The *Sv. Anna* became icebound shortly after passing through Yugorskiy Shar. She drifted northwards and never returned. A party left the ship in the summer of 1913 when she was north of Zemlya Frantsa-Iosifa [Franz Josef Land] and still icebound, and two members of this party succeeded in reaching land and safety. These were the sole survivors (370). The other voyage was planned by the Russian geologist V. A. Rusanov, who had led expeditions to Novaya Zemlya and Spitsbergen. Rusanov's object was to sail his ship *Gerkules* from the north end of Novaya Zemlya to the Pacific, calling at Ostrov Uyedineniya, Ostrova Novosibirskiye and Ostrov Vrangelya. After leaving Novaya Zemlya, Rusanov was never heard of again. Wreckage, evidently of the *Gerkules*, was found later on the west coast of Taymyr, so the assumption is that the ship came to grief in the Kara Sea (345).

In 1914–15 the second traverse of the whole length of the sea route, and the first from east to west, was made by B. A. Vil'kitskiy with the *Taymyr* and the *Vaygach*. This expedition was principally concerned with hydrography, and the results produced were valuable in this respect. But further, this was the first expedition whose sponsors—the Government—were thinking in terms of a possible through route from one end of the Russian Empire to the other. Their reasons were, as we have seen, strategic. The ships reached Arkhangel'sk, their destination, on their third attempt—ice in the vicinity of Mys Chelyuskina had made them turn back in 1912 and 1913. When they finally rounded Mys Chelyuskina in 1914 they were compelled to winter off the west coast of Taymyr before they could complete their voyage (10). It is possible that this success, qualified though it was, would have been followed up, since the hoped-for result was in the end obtained. But the time was unpropitious. In 1915 icebreakers were urgently required to keep open the White Sea for war supplies. Two years later the Revolution came and brought with it problems greater than that of the Northern Sea Route.

At this particular moment it so chanced that another through voyage from west to east was made. This voyage had nothing at all to do with the Russian Government or with commercial interests. The Norwegian explorer Roald Amundsen wanted to undertake a drift across the Arctic Ocean. In order that the course of drift should lie closer to the centre of the Arctic Ocean than the *Fram* had passed, it was necessary to start from a point farther east, nearer Bering Strait. Amundsen decided to reach his starting point by sailing to it along the north Siberian shore. He set out in his ship, the *Maud*, in 1918. Ice forced him to winter just east of Mys Chelyuskina, and the following year he again had to winter before reaching his objective, this time near Chaunskaya Guba in the East Siberian Sea. He reached Bering Strait in the summer of 1920. The drift, which ultimately took place from 1922 to 1924, only carried the ship back on its course to Ostrova Novosibirskiye (329). Both the voyage along the Siberian coast and the drift provided valuable scientific results which, like those of the *Fram* and *Zarya* expeditions, constituted essential background information of great importance to any subsequent expedition to these waters.

In view of the difficulties that Vil'kitskiy and Amundsen had had with the ice, the establishment of a workable sea route between Atlantic and Pacific did not seem to anyone in the years immediately following to be a particularly near possibility. It was doubted whether a ship could ever make the whole voyage in one season. It was fairly clear that the next step must be further study of those waters about which least was known—roughly speaking the central section between the Yenisey and the Kolyma—and of the behaviour of sea ice in general. During the 1920's there was a very considerable increase in the number of scientific expeditions to various parts of the Soviet Arctic. We will consider the growth of scientific work in the next section. One of these expeditions however must be mentioned here. In 1932 the *Sibiryakov*, carrying a party of scientists, completed the voyage from the Atlantic to the Pacific end of the Northern Sea Route in one summer. This voyage had far-reaching results, for its successful outcome was certainly one of the factors which decided the Soviet Government to embark on an extensive programme of arctic development.

The *Sibiryakov* was a small icebreaker of 1384 registered tons, launched in Glasgow in 1909 as the *Bellaventure*. The Arctic Institute acquired her for this expedition, whose object was to sail from Arkhangel'sk to Bering Strait, if possible in one season, and to carry out a big programme of scientific work *en route*. O. Yu. Shmidt, Director of the Arctic Institute, was leader. V. I. Voronin, an experienced ice navigator, was captain of the ship. V. Yu. Vize, Deputy Director of the Arctic Institute, was in charge of the scientific work. The voyage was successful, and far from uneventful. The ship sailed round the north end of Severnaya Zemlya—she was the first ship to do so—and lost a propeller blade in heavy ice as she came south again in the Laptev Sea. Off Kolyuchinskaya Guba heavy ice was met, one propeller blade was lost and the remaining three were badly damaged. By jettisoning much coal and food the stern was raised sufficiently to allow repairs to be carried out. The voyage was continued. On the next day the thrust-bearing went, and on the following day the propeller shaft broke and the screw fell to the bottom of the sea. The ship drifted towards Bering Strait in heavy ice, and with the aid of improvised sails she reached the neighbourhood of Mys Dezhneva where a trawler took her in tow. The voyage from the White Sea to Mys Dezhneva took 62 days [375].

It is worth noting that economic motives, which often exert a powerful influence on exploration, played little part in the discovery of the North East Passage. Only the earliest attempts were prompted by the hope of financial gain. The successful voyage of the *Vega* was inspired by no such motive. After that voyage it was the demands of science which took ships to the waters traversed by the *Vega*. Thirty years passed before the idea of making a sea route out of the North East Passage was again considered; and then the motives were strategic rather than economic. The difficulties encountered in 1912–15 damped the enthusiasm of any who may have wanted to follow up the partial success achieved at that time; and for the next fifteen years interest in the area was again limited to scientific expeditions working on tasks not directly concerned with the establishment of a sea route.

5. SCIENTIFIC WORK

It became clear not long after the first voyages to the Ob' and Yenisey that if shipping was to be made safe in the Kara Sea, a hydrographic survey of the shipping lanes must be undertaken. In the case of the Chukchi Sea and the East Siberian Sea this was realised when the idea of a sea route to the Kolyma was first scouted. As time went on, and especially as ships started to go farther afield in these arctic seas, it became clear that much more than the hydrography of the region required to be known. It was necessary in addition for the navigator to have information on sea ice and its behaviour, on local weather conditions, and on terrestrial magnetism. If wireless communication was to be secure, the behaviour and propagation of radio waves in high latitudes must be studied. In fact an enormous amount of work was waiting to be done in a number of branches of science before ships could sail confidently in these waters.

An outline has already been given of the hydrographic work which was performed in north Siberian waters before the Revolution. Broadly speaking, the work was done by the Hydrographic Expedition of the Arctic Ocean in the Kara Sea between 1894 and 1904; and by the series of expeditions in the

Taymyr and the *Vaygach* between 1910 and 1915. On the face of it, there was cohesion and plan here, since all the expeditions were organised by the same authority, the Hydrographic Administration [Gidrograficheskoye Upravleniye] of the Ministry of Marine. To a limited extent this was so. But it may be noted that the work in the Kara Sea from 1894 to 1904 was started twenty years after the need for it was first felt; and the next phase, which had been planned to continue in 1910 where work had been left off in 1904, in fact started at the opposite end of the Russian Arctic, and then continued there for at least two years longer than was envisaged by the revised plan because ice prevented the *Taymyr* and the *Vaygach* from coming back to the west. Most of the other scientific work was done in a very haphazard way, as might indeed have been expected, reflecting the spasmodic development of the shipping routes. In the field of meteorology and the study of sea ice important work was done by the stations set up for the International Polar Year of 1882–83. This was an international scheme to increase knowledge of the polar regions by the simultaneous recording of scientific information at various selected points. The Russian Government manned stations at Malyye Karmakuly on the west coast of Novaya Zemlya, and at Sagastyr' at the mouth of the Lena. Dutch and Danish parties were to work at Ostrov Diksona and Mys Chelyuskina, but their ships were caught in the ice and drifted all winter in the Kara Sea. Nevertheless the drifting parties as well as those on shore recorded meteorological information which was of particular value because all the instruments used had been tested beforehand. The station at Malyye Karmakuly was manned, with occasional short gaps, from 1896. Later, an important contribution was made by the polar stations established from 1913 on the shores of the Kara Sea and its approaches. It is relevant to mention the investigations conducted in the Barents Sea from 1898 to 1908 by the Murman Scientific-Industrial Expedition [Murmanskaya Nauchno-Promyslovaya Exspeditsiya]. The principal object of these investigations was aid to fisheries, but the knowledge gained of behaviour of the Barents Sea was clearly relevant to study of waters farther east. The same is true to a lesser extent of the Murman Biological Station [Murmanskaya Biologicheskaya Stantsiya], which worked on marine biology in coastal waters from 1904 to 1933, with a gap during the war years.

One of the principal effects of the Revolution on the development of north Russian waters for shipping was a gradual and considerable expansion of scientific work and the creation of new institutions to undertake this work. It was the western sector which was chiefly affected. Hydrographers and hydrologists worked in the estuaries of the Ob' and Yenisey, off the east coast of Novaya Zemlya, in the entrance to the White Sea and off the Murman coast (273). New sailing directions were issued for Yeniseyskiy Zaliv in 1924 and Obskaya Guba in 1925. From about 1929 the range of action was extended and parties worked in the vicinity of Zemlya Frantsa-Iosifa and in the northern and eastern parts of the Kara Sea (131), including the rocky and dangerous west coast of Taymyr (108,140). A preliminary study of the river Pyasina was made in 1932 with the principal object of finding a safe channel into the river from the sea (45). In the eastern seas however, after the conclusion of the *Taymyr* and *Vaygach* expedition in 1915 no more hydrographic work was done on a comparable scale until after 1932. Some Soviet ships took soundings near Ostrov Vrangelya in 1924 and 1926, and in 1932 a hydrological expedition worked in Bering Strait (259). But the rivers of eastern Siberia were the subject of considerable attention. A detailed survey of the Lena below Yakutsk, including the delta, was made between 1919 and 1921. Work on the Kolyma in 1928–29 resulted in the publication of an atlas of the river. Another expedition worked on the Indigirka in 1931 and an atlas of this river was produced. A detachment of the Indigirka expedition explored the Ozhogina, a tributary of the Kolyma. The Yana was navigated as far as Verkhoyansk in 1927–29 (80).

All the sea hydrographic work was undertaken by the Hydrographic Administration of the Soviet Navy [Gidrograficheskoye Upravleniye VMF] which formed a special department to deal with the Kara Sea. The expeditions to the east Siberian rivers were sponsored by the People's Commissariat for Water Transport [Narodnyy Komissariat Vodnogo Transporta, abbreviated to Narkomvod]. Komseverput' and the State Hydrological Institute [Gosudarstvennyy Gidrologicheskiy Institut] each contributed on a smaller scale. The Arctic Institute[1] performed some of the hydrological work. This phase of hydrographic and hydrological work compared favourably with the pre-Revolutionary phase. The volume of work was greater, and in the Kara Sea at least some sort of continuity was evident. But there was too little co-ordination between the various organisations, and there was no overall plan at all. Hydrological work in particular continued to be performed by members of expeditions of which the principal object was generally geographical. In the early stages this did not matter particularly, since there was work to be done wherever any expedition might go; but once a certain minimum had been done it clearly became necessary to plan future work if the best use was to be made of the knowledge gained. This stage was reached in the western sector about 1930, but no plan was yet forthcoming. In the eastern sector the basic minimum remained to be done.

A considerable amount of oceanographical work was carried out in the west, chiefly on behalf of fisheries. On an average three or four expeditions a year worked in the Barents Sea, with occasional excursions into the south-western part of the Kara Sea, between 1921 and 1932. The scientists accompanying these expeditions included zoologists and marine biologists as well as hydrologists. Very much stronger financial support made these expeditions more effective than their predecessors of 1898–1908. This series of expeditions was organised by the Floating Marine Scientific Institute [Plovuchiy Morskoy Nauchnyy Institut] which later was fused with the State Oceanographical Institute [Gosudarstvennyy Okeanograficheskiy Institut].

The usefulness of the icebreaker to scientific work was realised. In 1928 the Soviet Government sent two icebreakers, the *Krasin* and the *Malygin*, in search of the survivors from the Italian airship *Italia* which had crashed

[1] This Institute's frequent changes of name are omitted here, but are given in the footnote on p. 98. In this study the Institute will always be referred to as the Arctic Institute.

north-east of Svalbard on its way back from the North Pole. The *Sedov*, another icebreaker at the time on a hunting voyage, later joined in the search. The survivors were found. The important point is that the *Krasin* and the *Malygin* expeditions were each led by a scientist, and each ship carried a group of scientific workers who took hydrological and meteorological observations *en route* (272). These voyages demonstrated the possibilities of the icebreaker as a ship for a scientific expedition. Further voyages were organised, with scientific work as a principal and not subsidiary object. These account for the extended range of the hydrographic and hydrological work after 1929, noted above. In 1929 (377) and 1930 (46) the *Sedov* carried parties of scientists to Zemlya Frantsa-Iosifa and Severnaya Zemlya. Four men were left on Severnaya Zemlya in 1930, and during the next two years they mapped the whole of this group of islands which constituted the last considerable unexplored area north of the Eurasian mainland. In 1931 the *Malygin* (378) went to Zemlya Frantsa-Iosifa, which she visited again twice in 1932 (206). Also in 1932 the *Rusanov* (271) cruised in the north-east part of the Kara Sea, and the *Sibiryakov* (375) made her famous traverse of the Northern Sea Route. On all these voyages a variety of scientific work was done, with the emphasis on hydrology, meteorology and study of ice conditions. The icebreaker escorting merchantmen in the Kara Sea also carried a hydrologist. These icebreaker expeditions were all organised by the Arctic Institute.

Occasion was taken on several of the later voyages to land and establish polar stations at various points. In this and other ways nineteen coastal stations were set up by the end of 1932, in addition to the five which had existed at the time of the Revolution.[1] There was also a growing number of inland meteorological stations; these helped to provide an overall picture of weather conditions but do not concern us directly. Of the 24 stations, all but six served the Barents Sea or the Kara Sea. These two seas were thus quite well supplied with stations, while the seas farther east still had only a skeleton network. Construction of and control over polar stations was the concern of a number of organisations, notably the meteorological authorities and the Hydrographic Administration. The Arctic Institute set up its first station in 1929, and was responsible for many of those erected after that date.

Mention must be made of the Second International Polar Year of 1932–33. This event, like its predecessor of 1882–83, was organised with the object of obtaining the simultaneous recording of certain geophysical observations at many points. The Soviet Government decided to include in the scheme a large number of polar stations and mobile scientific expeditions in ships. Some new polar stations were set up specially for this event, under the control of the Arctic Institute. Some of these were evacuated in 1933, but the others became permanent stations (only the permanent stations are included in the total of 24 above).

The work done by polar stations continued to be chiefly meteorological. The regular recordings made at a growing number of points over a period of years assumed increasing importance to weather forecasters. From 1925 a

[1] See Appendix IX for fuller details of polar stations.

forecasting team of two men worked on board the leading ships or icebreaker of the Kara expedition, collating and disseminating the daily reports of stations in the area. In 1932 another worked in the *Litke*, escorting the Kolyma convoy (44). Observations of sea ice and tides were made at most stations. In addition, a start was made on magnetic work. Magnetic observatories were established at Matochkin Shar in 1923, at Bukhta Tikhaya in 1931 and at Ostrov Diksona in 1932.

On the basis of hydrological and meteorological work done, the first steps were taken in a new field—ice forecasting. A forecast in general terms of ice conditions in the Barents Sea during the coming summer was issued in 1923 by the Central Hydrological and Meteorological Bureau [Tsentral'noye Gidrometeorologicheskoye Byuro]. A forecast was made each year thereafter and the Kara Sea was later included. For the period 1923–28 45 % of these forecasts were estimated to have been "wholly successful" and 35 % "satisfactory". The State Hydrological Institute took over the service in 1929, and in 1932 forecasts were for the first time issued for all five seas in view of the impending voyage of the *Sibiryakov*. Special forecasts were also made for the Kara Sea traffic by the Chief Geophysical Observatory [Glavnaya Geofizicheskaya Observatoriya] (368).

The establishment and work of polar stations in this period bear the same characteristics as were apparent in the case of the hydrographic and hydrological work: while the volume of work was very considerably increased after the Revolution, little if anything was done to co-ordinate it. But the desire to increase scientific knowledge of the Arctic regions was clearly present, both in the Government and in the scientists. In these circumstances it could only be a matter of time before the necessary co-ordination was inaugurated from above.

PART II

THE NORTHERN SEA ROUTE
FROM 1933 TO 1949

From 1933 the use of Soviet Arctic waters by shipping greatly increased, as a result of the decision of the Government to embark on an extensive programme of Arctic development. The possible motives for this decision we shall discuss later, when we consider the uses to which the route was put. It will suffice here to mention that according to V. Ya. Chubar', a deputy chairman of the Council of People's Commissars (equivalent to our Cabinet), Stalin put the problem in this way: "The Arctic and our northern regions have colossal wealth. We must create a Soviet organisation which can in the shortest period include this wealth in the general resources of our socialist structure" (196). Assuming for the moment that this was the principal motive behind the Government's action (and there is good reason to believe it was, as we shall see later), we may ask why the action should have been taken at this particular time. In fact it would have been inopportune to embark on so large a project earlier, since in the 1920's the Government was struggling with problems of more pressing importance—recovery from war and civil war, famine, social upheaval caused by the collectivisation of the peasants. The end of 1932 was also the end of the first five-year plan, of which the principal achievement was the establishment of a heavy industry as an essential preliminary to wider industrialisation. This was a suitable time to launch a large-scale programme of development, since it could be integrated with the national economy through the second five-year plan. The successful voyage of the *Sibiryakov* may be regarded as an important contributory factor since it made clear that technical skill and equipment were in fact sufficiently advanced to permit the opening up of hitherto unexploited parts of the Arctic coast.

In implementation of the decision all sorts of things were done. A vast amount of effort and money was expended, and the Northern Sea Route, which was always regarded as the cornerstone of northern development, received the largest share. The principal feature of this communist attack upon the Arctic was to be its careful planning, which would distinguish it from the haphazard and sporadic activity before the Revolution. Bearing this point in mind, we will go on to consider what was done and examine its effectiveness.

1. SOME PHYSICAL CHARACTERISTICS OF THE ROUTE

The events which we shall be examining should be seen against a background of the physical conditions in which they took place. It is not the intention of this study to give a detailed geographical description of the whole Northern Sea Route because such descriptions, in English, are available elsewhere.

Nevertheless some information on the principal features affecting navigation should find a place here.

(i) *Routes used*

The main shipping lane remained virtually the course of the *Vega* in 1878. East of the Kara Sea the route lies within fifty miles of the shore almost the whole way, leaving coastal waters only to cross the larger indentations of the Laptev Sea. The Kara Sea is entered, as before, by Yugorskiy Shar, Karskiye Vorota, Matochkin Shar or the northern end of Novaya Zemlya. Proliv Vil'kitskogo is the normal communicating channel with the Laptev Sea; Proliv Shokal'skogo is seldom used, being frequently icebound. East of the Laptev Sea both Proliv Lapteva and Proliv Sannikova are used; and entry to the Chukchi Sea is normally by Proliv Longa.

In 1932 the *Sibiryakov* found impassable ice in Proliv Vil'kitskogo, and was able to avoid it by passing round the northern end of Severnaya Zemlya. This feat was not repeated certainly until the war, but it led to frequent discussion of the possibility of developing a so-called "northern variant" of the main route, passing north of Novaya Zemlya, Severnaya Zemlya, Ostrova Novosibirskiye and Ostrov Vrangelya. Various icebreaker voyages to high latitudes provided information on ice conditions in these waters (see p. 89). During the third five-year plan (1938–42) serious attention was given to the northern variant, and it was proposed to build polar stations on the north shores of the island groups (see p. 91). In 1940 and 1941 the Arctic Institute and the polar aviation organisation were given the task of studying the northern variant (233). Lack of more exact knowledge about it was deplored in 1940 when apparently ships could have passed north of Ostrov Vrangelya (284) but no aircraft were available to reconnoitre the route onwards. To what extent the northern variant was further developed or used during and after the war is not known. It can only be said that considerable thought was given to the matter. But it may be added that at least one ice expert, A. F. Laktionov, is sceptical about the possibility of making regular use of the route north of Severnaya Zemlya. He (132) writes, probably during the war: "The use of this variant as a route from the Kara to the Laptev Sea when ice conditions are bad in the southern straits is scarcely possible. A single circumnavigation of Severnaya Zemlya, which took place in 1932, is insufficient to resolve the question of the possibility of using this route even in good ice years."

(ii) *Sea ice*

Sea ice is of course the principal hindrance to shipping. It is impossible, however, to give any fixed limits to the duration of the navigation season, since the dates of break-up and freeze-up vary so greatly from year to year. It can only be said that the season lasts for anything from 70 to 120 days between the end of June and the middle of November. During this period parts of the route are open; the whole route is of course open for a shorter period than its parts.

The rivers are always clear of ice before the sea because their upper reaches in more southerly latitudes thaw earliest and the pressure of ice and melt water coming downstream accelerates the break-up on the lower reaches. The table below (35) shows dates of clearance and freeze-up.

Place	Free of ice			Freeze-up		
	Earliest	Latest	Years of observation	Earliest	Latest	Years of observation
Ob' at Novyy Port	10 June	3 July	1925–33	12 Oct.	25 Nov.	1924–33
Yenisey at Dudinka	26 May	17 June	1741, 1900, 1903, 1905–07, 1932, 1934–36	22 Oct.	3 Nov.	1843, 1904–07, 1932, 1934–37
Lena at Bulun	27 May	7 June	1909–17, 1921, 1927–36	10 Oct.	31 Oct.	1909–16, 1926–35
Yana at Kazach'ye	23 May	18 June	1823, 1869–71, 1873–76, 1879–81, 1883–90, 1892–93, 1896, 1899, 1901–03, 1920–28	10 Sept.	9 Oct.	1821–23, 1875, 1884–85, 1895–96, 1902–03, 1920, 1928
Indigirka at Russkoye Ust'ye	11 June	20 June	1895–1903, 1905	16 Sept.	10 Oct.	1821, 1896–1903
Kolyma at Nizhne-kolymsk	27 May	17 June	1787, 1810, 1821–24, 1869, 1880–82, 1892–93, 1897, 1900–05, 1909, 1912, 1928–29	20 Sept.	20 Oct.	1787, 1810, 1821–23, 1880–82, 1892–97, 1900–04, 1928

There is reason to suppose that since about 1920 the Arctic regions have been getting warmer. Evidence of this trend has been collected in many parts of the Arctic: retreat of glaciers in Scandinavia, higher air temperatures in Greenland, Svalbard and the Soviet Arctic, northwards movement of cod in the North Atlantic, to mention only a few instances. It has been noticed that the sea ice has been affected by this process. The higher air and water temperatures have some effect, but the increased storminess, itself a result of the process, has a greater effect by breaking up the ice and delaying its freezing. In 1901 the icebreaker *Yermak* was unable to force her way round the northern end of Novaya Zemlya, while since 1930 ordinary freighters have been able to sail round without difficulty every year (404).

Vize (367) calculates that the likelihood of encountering ice in September in the Kara Sea south of the parallel of Matochkin Shar was 30% during the years from 1869 to 1928, while from 1929 to 1939 no ice was seen there in September. The same process is thought to be going on in the eastern sector also (404), but figures illustrating decrease in the amount of sea ice are not available. The results of two drifting expeditions in the central polar basin show (403) that the air temperatures in 1937–39 were on an average 4° C. (7·2° F.) higher than those in 1893–95; during the six winter months they were 7·5° C. (13·5° F.) higher. The speed at which the ice drifts across the basin was also observed to have increased. The warming up of the Arctic and the effect of this on sea ice may well have been a factor in the general success of the shipping seasons of the 1930's.

(iii) *Fogs and storms*

In the limited period during which navigation is possible there are two further obstacles: fogs and storms.

Fogs are the most serious hindrance, since the peak period for fogs coincides largely with the navigation season. September is generally better than August however, and October is better still. The table below (122) shows the frequency of fogs at various points.

Average number of days per month with fog (excluding sea and ground fog)

Place	Years of observation	June	July	August	Sept.	Oct.
Yugorskiy Shar	1913–35	15	17	17	10	7
Matochkin Shar	1924–35	8	8	10	10	5
Mys Zhelaniya	1931–35	14	18	23	13	4
Ostrov Diksona	1916–35	14	19	15	12	5
Novyy Port	1924–33	8	6	7	7	7
Ust'-Port	1921–33	6	2	4	5	7
Mys Chelyuskina	1932–35	14	24	25	16	8
Ostrova Komsomol'skoy Pravdy	1933–35	17	17	22	10	2
Bukhta Tiksi	1932–35	15	10	5	5	4
Mys Shalaurova	1928–35	15	17	16	6	4
Ostrov Chetyrekhstolbovoy	1933–36	16	21	23	10	3
Mys Shelagskiy	1934–36	21	17	20	12	2
Mys Shmidta	1932–36	13	16	18	12	3
Ostrov Vrangelya	1926–36	15	17	16	8	4
Mys Uelen	1928–36	14	16	14	12	4

The incidence of storms is not nearly so serious. The winds along most of the Northern Sea Route exhibit a monsoonal character, blowing landwards from the sea from May to August and in the reverse direction in the winter. East of the Indigirka however the Aleutian minimum plays a part, and northerly inshore winds tend to blow for ten months of the year, with southerly winds only in July and August (366). The table below (366) shows the incidence of storms:

Average number of days per month with storms (wind speed greater than 16 metres per second)

Place	June	July	August	Sept.	Oct.
Yugorskiy Shar	3	2	2	4	8
Ostrov Diksona	3	3	4	4	7
Ostrov Vrangelya	1	3	2	5	5

2. TRAFFIC BORNE BY THE NORTHERN SEA ROUTE

(i) *Sea shipping*

All the movements of shipping anywhere along the Northern Sea Route were from 1933 carried out under the aegis of the Chief Administration of the Northern Sea Route [Glavnoye Upravleniye Severnogo Morskogo Puti, abbreviated to Glavsevmorput'], the new department specially created to carry through the Government's programme in the Arctic. The fact that there was unified control made it possible for a detailed plan to be worked out for each navigation season. Reference to under- or over-fulfilment of plans is an accepted Soviet way of judging any undertaking, and will therefore appear below from time to time. It is not a very satisfactory way from our point of view since detailed knowledge of the plan itself is almost always lacking; but it is better than nothing.

The voyages made fall into various groups: the voyages along the whole length of the route and other long-range voyages; voyages to the Lena; and voyages on the already established routes to the Kolyma and to the Ob' and Yenisey—these came to be known as the Kolyma operations and the Kara operations. A general outline of these shipping movements, year by year, is given below.[1]

1933

A through voyage was planned for a new ship, the *Chelyuskin*. This ship was of a new type, being a freighter with her hull strengthened for ice and including various other features of an icebreaker in her design. Great hopes were placed in her, for she was regarded as the prototype of the future Arctic freighter. She was to make the voyage unescorted, and the leader of the expedition was O. Yu. Shmidt, the Head of Glavsevmorput'. There were difficulties with the ice in the Kara Sea, where she spent eighteen days and sustained some damage before reaching Proliv Vil'kitskogo. She continued the voyage, met heavy ice off the north coast of Chukotka and finally was trapped by the ice almost in the entrance to Bering Strait. She was unable to free herself, and in February 1934 she was crushed by pressure of ice and sank. Those on board were rescued by a remarkable series of flights to the sea ice where a camp had been set up (288). This loss was a severe one for Glavsevmorput' in its early days.

The year 1933 marked the opening of a new shipping route, from the west to the Lena. The object of the route was to provide a new link between central and northern Yakutiya and the outside world. There is not sufficient water to allow ocean-going vessels to enter the Lena. It was necessary therefore in the first place to choose a suitable trans-shipment point where the river craft could meet the ships. The place chosen was Bukhta Tiksi, to the south of the entrance to the main easterly channel of the Lena delta. The first convoy to go to the Lena, called the first Lena expedition, consisted of three freighters sailing from

---

[1] Available details are given in tabular form in Appendices I–III: Appendix I, Kara operations; Appendix II, Kolyma operations; Appendix III, voyages to the Lena. The total yearly freight turnover of the whole Northern Sea Route is given in Appendix IV.

Arkhangel'sk, a river tug and barge from the Ob', and an escorting icebreaker, the *Krasin*. All but one of the freighters went to Tiksi and deposited cargo, chiefly port equipment. The tug and barge stayed on the river. The remaining freighter went to Nordvik, at the mouth of the Khatanga river, but shallow water in the roads compelled the ship to unload at Bukhta Pronchishchevoy some 120 miles to the north. The three freighters reassembled for the homeward journey but were caught in the ice off the east coast of Taymyr and were compelled to winter there (147).

Traffic continued on the Kolyma route. It will be remembered that the large expedition of 1932—the icebreaker *Litke* and six freighters—was forced to winter at sea off the north coast of Chukotka. These ships were freed in July 1933 after the ice had broken up. Four of them, with the *Litke*, went back to the Kolyma again to continue unloading stores that they had been unable to unload the previous year. Two of these four were caught in the ice for the winter a second time. The *Litke* and the remainder got back to Vladivostok. Meanwhile four ships and a schooner came up from Vladivostok with goods for the Kolyma and the coastal polar stations in the area. One of these was also compelled to winter at sea, while the rest returned to the Pacific (390).

The Kara Sea saw a continuation of the previous year's traffic to Ob' and Yenisey. Thirty merchantmen were employed; 25 went to Igarka and five to Novyy Port. One ship went aground off Vaygach on the return journey and had to be abandoned, but the others had uneventful voyages (280).

1934

The icebreaker *Litke* made the first traverse of the Northern Sea Route from east to west in one season. The main objects of the voyage were to make this traverse and to help shipping in the Laptev Sea. A scientific party was on board. The voyage was successful in both its objects, though the *Litke* sustained some damage from ice when she was freeing the ships of the 1933 Lena expedition in the Laptev Sea, and had to be freed from the ice herself in the Kara Sea. She remained in the vicinity of Ostrov Diksona for nearly two weeks in order to help ships crossing the Kara Sea from the Yenisey (381).

Three ships escorted by the icebreaker *Yermak* formed the second Lena expedition. The convoy, sailing from Murmansk, reached the Lena and returned westwards without mishap (106).

The Kolyma expedition was made up of three ships, whose voyage was also uneventful. Supplies were offloaded *en route* at points along the coast as well as at the Kolyma itself (274).

Twenty-eight ships sailed to the Ob' and Yenisey from the west; 23 went to Igarka, five to Novyy Port. The three ships of the first Lena expedition of 1933 called at Igarka after they had been freed from their wintering place. The icebreaker *Malygin* was allotted to the Kara Sea. The Kara expedition was run jointly with the Lena expedition this year, but the latter had a separate icebreaker, the *Yermak*, at its disposal (281).

The total number of ships employed is given by one source as 85 (227). This figure is much larger than the total of the operations listed above, but includes ships carrying scientific expeditions or relief parties and others doing relatively unimportant jobs.

1935

Now that the loss of the *Chelyuskin* had been to a certain extent offset by the successful voyage of the *Litke* in 1934, the way was considered clear for another attempt at a through voyage by freighters. This time the attempt was successful. Two ships made the traverse in one direction, and two more in the other. They were helped by icebreakers when necessary (43).

Two other exceptional voyages were made, both the first of their kind: one cargo vessel sailed from the west to the Kolyma and returned the same season; another made a similar voyage to the Indigirka and back (123).

A new system of shipping control was introduced in 1935. In place of overall control by one man—the Head of Glavsevmorput' or his deputy—the area of navigation was divided into an eastern sector and a western sector, with Mys Chelyuskina as the dividing point. Each sector had a Director of Operations who was on board the duty icebreaker in that sector. In 1935 the *Yermak* performed that function in the west, with the *Lenin* under command and in charge of the Kara operations. The *Krasin* was the flagship for the east, assisted by the *Litke* which was responsible for the Laptev Sea (43, 287). The system seems to have worked well. It was an improvement on the system of control by one man, who could not have as detailed a picture of conditions as each Director of Operations had.

The Lena operation consisted of five freighters which had an uneventful voyage (239).

Five ships sailed to the Kolyma from Vladivostok. Two of these continued their voyage westwards after discharging, and these were the two ships which made the east to west traverse. One ship, as we have already remarked, came to the Kolyma from the west. In addition, three ships supplied the settlements and polar stations of the eastern sector (239).

Kara Sea operations employed 45 ships. All but five went to Igarka. Nine came from points farther east along the route: one of these came from Vladivostok, called at the Kolyma and at Igarka, and then went straight through to London—a new demonstration of the potentialities of the route (282).

Taken as a whole, 1935 was undoubtedly a successful year. It was officially considered the year of "trial exploitation", and the Head of Glavsevmorput' expressed satisfaction at the results. The total turnover was said to be 113 % of the plan (292). An editorial in the journal of Glavsevmorput' proudly claimed that "the Northern Sea Route has become a normally working route" (197).

1936

The successes of 1935 led to the adoption of an ambitious plan for 1936, but bad ice conditions in August interfered to a considerable extent with its successful fulfilment. The two-sector system was retained, and again two icebreakers were allotted to each sector. In fact, conditions made necessary the diversion of several more icebreakers from scientific work.

A convoy of sixteen merchantmen, made up of five bound for the Lena, three for the Kolyma and eight for the through voyage, was escorted eastwards across the Kara Sea by five icebreakers. Even with so much assistance as this, none of the ships reached the Laptev Sea until early September. Those bound for Tiksi discharged and returned safely. The ships for the Kolyma were still unloading on 1 October. It was clearly out of the question at this stage in the season to send them back to the west, as had been planned. They were therefore escorted eastwards. The through voyage from west to east was made by twelve ships: eight as intended—two of these were only powered sailing boats of 135 tons each—and the three Kolyma ships with their escorting icebreaker. From east to west two ships completed the traverse, calling at the Kolyma, Tiksi and Nordvik. Three ships bound for Nordvik from the west were forced, through lateness in the season and other causes, to abandon their voyages when still in the Kara Sea (49).

The Kolyma was thus supplied by three ships on the west to east traverse, and two ships on the east to west traverse. In addition one ship from Vladivostok took freight there, went on to Tiksi and Nordvik, and returned to Vladivostok. Four ships supplied the coastal settlements east of the Kolyma (49).

The Kara Sea route operated as usual, on the same scale as the previous year, with the icebreaker *Malygin* in charge (291).

Considering the unfavourable ice conditions, the results of the season in terms of freight turnover were good—there was an increase of 18 % on the 1935 total (291). It was now becoming clear however that sea transport had developed too quickly for many of the shore installations. The principal ports—Ostrov Diksona, Bukhta Tiksi and Bukhta Provideniya—were little more than anchorages, in most cases without wharves. Coaling facilities were inadequate. If seaborne traffic was to go on increasing at the same rate, much leeway would have to be made up.

1937

It was planned that four ships should make through voyages, and one of them was to make the double voyage from Murmansk to Kamchatka and back. The usual Lena, Kolyma and Kara operations were planned and some ships were to call at Nordvik. The Kolyma operations were evidently to consist chiefly of ships coming from the west. The ships bound for Nordvik, the Lena and the Kolyma from the west left the Kara Sea under escort in several convoys. Not all of these convoys succeeded in reaching their destinations owing to bad ice conditions in the Laptev Sea and the eastern Kara Sea. No information is available as to the number of ships employed, but four-fifths of the Kolyma ships, two-fifths of the Lena ships and a third of the Nordvik ships were all that succeeded in unloading at their specified destinations (4). Again owing to ice, some of these ships continued eastwards instead of returning to the port they had come from. In this way six ships made through voyages "by mistake". The total number of west to east through voyages was ten, of east to west, one. The double voyage failed by a narrow margin; the ship had to winter at sea near Mys Chelyuskina on the return trip. The freight programme as a whole was thus considerably under-fulfilled. Only just over three-fifths of all freighters (apart from those taking part in the Kara operations) were able to unload at their appointed places.

But there was a much worse disaster at the end of the season. A convoy of five ships bound for Tiksi under escort by the icebreaker *Lenin* became icebound near Proliv Vil'kitskogo at the end of August, and started to drift through the straits into the Laptev Sea. In the middle of September the icebreaker *Litke* and another convoy of five, westbound from Tiksi, could not penetrate the ice at the entrance to Proliv Vil'kitskogo. The *Krasin* tried to help the *Lenin* but got trapped herself. Three small icebreakers, the *Sedov*, *Sadko* and *Malygin*, were diverted from scientific work to help ships in the Laptev Sea, and all three were caught by the ice west of Ostrova Novosibirskiye in October while they were trying to escape eastwards from the impassable ice blocking Proliv Vil'kitskogo. The *Yermak* made valiant efforts to release what ships she could of the large number which were helpless, but she failed and only with difficulty escaped herself. The remaining icebreaker left in service, the *Rusanov*, freed herself after a month from the ice in the Kara Sea where she had been relieving polar stations, and was directed to Zemlya Frantsa-Iosifa [Franz Josef Land] to help two ships already icebound there. It was very late in the season, and it is scarcely surprising that in the end all three ships had to winter at sea (4, 50).

The only phase of the planned shipping movements which went comparatively normally in 1937 was the Kara operation. This was planned on the same scale as the year before. It was necessary however to send out two extra convoys to Igarka from Arkhangel'sk late in September, when it became clear that ships returning from points east of the Yenisey would not be able to free themselves and collect cargoes at Igarka. The first of these late convoys returned safely at the end of October, but the second, consisting of six ships, was caught by the ice in the vicinity of Ostrov Diksona (50).

Thus at the end of the season 26 ships were wintering at sea; among them seven out of the eight serviceable icebreakers owned by Glavsevmorput'. There had been no catastrophe like this at any time since ships first started to sail in these waters. There were violent repercussions in almost every department of Glavsevmorput', and wild political accusations were made in attempts to allot blame. We will consider the far-reaching consequences when we come to deal with the structure of Glavsevmorput'. It will be appropriate here to estimate, as far as possible, the causes of the disaster.

There is little doubt that it was not ice conditions alone which were to blame. In many places and for much of the season the conditions were favourable. This is illustrated by the very successful outward voyage of the ship attempting the double traverse. The ships were not always in the right place at the right time to make use of the good conditions. This was certainly due, in part at least, to bad organisation. All analysers (4, 54, 402) of the disaster agree that the absence of sufficient aircraft for ice reconnaissance was one of the chief disruptive factors. During the summer of 1937 there were some spectacular Soviet expeditions to and over the North Pole. Three trans-polar flights were made, and a party was conveyed by air to the North Pole and established on the ice there. In order to service these expeditions the majority of the aircraft normally used for ice reconnaissance and aid to shipping were diverted to Zemlya Frantsa-Iosifa. The lack of aircraft on the Northern Sea Route—only two were in use—resulted in unreliable forecasts of ice and weather conditions and consequently in ineffective placing of icebreakers. Slow and inefficient transmission of certain wireless messages aggravated the situation. It seems clear also that the coal used for bunkers at Tiksi had much to do with the difficulties of the ships which refuelled there. The coal, which was mined at Sangar-Khaya on the Lena, was evidently of poor quality and as a result ships of the *Litke*'s convoy found they could not raise enough power to force the ice (397). Other reasons too were found for the catastrophe: some of the ships used had been certified as unfit for navigation in ice; and the division into two sectors was attacked on the grounds that there was no one to arbitrate between the two Directors of Operations. Further, and perhaps most important of all, over-confidence seems to have been the prevailing mood. The authorities at Murmansk even refused a ship-master's request for emergency winter rations and clothing on the grounds that the Arctic was already conquered and wintering at sea was no longer a possibility (402). Thus although the state of the ice in the Laptev Sea and Proliv Vil'kitskogo was undoubtedly a major hindrance, the human factor caused a sharp deterioration in a potentially bad situation.

1938

The first necessity was to extricate the ships which had wintered at sea. The only icebreaker available for this was the *Yermak*. She left Murmansk in May and relieved the ships at Zemlya Frantsa-Iosifa. On her next voyage she released the Kara operation ships near Ostrov Diksona, and went on to Proliv Vil'kitskogo, where she conducted the *Litke*'s convoy into clear water. Late in August the *Yermak* reached the group of three icebreakers, *Sedov*, *Sadko* and *Malygin*, which were drifting in the ice in the north part of the Laptev Sea. Two of the group were released but the *Sedov*'s rudder became damaged and it was decided to leave her behind as her slow speed endangered the escape of the others. The *Sedov* in fact drifted across the central polar basin for another year and a half, and returned to service in 1940. The *Yermak* thus reached four out of the five groups of wintering ships, and released all but one of the ships. For this work of retrieving, nearly single-handed, the disaster of 1937, the crew were awarded a special money prize. There remained the convoy of the *Lenin* off the east coast of Taymyr. One of the ships of this convoy, the *Rabochiy*, was crushed by ice pressure during the winter and sank. Her crew and some cargo was taken aboard the *Lenin*. The convoy was released by the *Krasin*, which had wintered near Nordvik and had been able to get enough coal locally to allow her to move off under her own power in the summer (51).

Much time and energy that would normally have been devoted to the passage of cargo ships was thus spent in relief work. The result was a smaller freight turnover than in 1936. A convoy from the west reached its destinations in the Laptev Sea—Nordvik and Tiksi—and returned safely. The indefatigable *Yermak* escorted it through the difficult stretch near Proliv Vil'kitskogo. In the eastern sector two ships made the voyage from Vladivostok to the Kolyma and back; two tankers went to Tiksi and back; and seven ships took supplies and relief personnel to polar stations and settlements. There was no icebreaker available in these waters until the *Krasin* released herself.

One through voyage was made by a tanker which sailed from east to west calling at Tiksi, Nordvik and Murmansk.

Kara operations were continued, with 45 ships calling at Igarka (51).

The first Soviet-built icebreaker, the *Stalin*, came into use in 1938. She did not reach Arctic waters until late in August. All she was able to do was to make an unsuccessful attempt to reach the *Sedov* after the *Yermak* had left her behind. The *Kaganovich*, the *Stalin*'s sister ship, which was due to come into service this year and work in the eastern sector, was not completed in time.

1939

In 1939 both the *Stalin* and the *Kaganovich* were available for service. There is little detailed information available on the navigation season, but some indication of the number of ships employed is given by the statement (406) that during the season 114 sea-going vessels called at Ostrov Diksona, twenty at Tiksi and 42 at Bukhta Provideniya.

Ten ships made through voyages from west to east. This total includes a group of four dredgers and four tugs, which had icebreaker escort only at difficult places (277). Another of the ships making a through voyage was the *Stalin*, which also succeeded in returning in the same year and was thus the first ship to make the double traverse.

Convoys left the western termini bound as usual for the Khatanga, Lena and Kolyma. For the first time a ship sailed from the west direct to the Yana and back. Hitherto the Yana had only been reached by coastwise traffic principally from the Lena. In the eastern sector there were the usual voyages to the Kolyma and elsewhere (396).

The Kara operations continued on a larger scale than ever. Novyy Port on the Ob' estuary had by now practically ceased to be used, while on the Yenisey Dudinka came into use as well as Igarka (179).

The navigation season of 1939 was called "the first year of normal commercial exploitation". The significance of this epithet was that for the first time the question of cost was taken seriously, and an attempt was made at least to explore the possibility of making the Northern Sea Route a paying proposition. The season was considered by the Head of Glavsevmorput' to have been a success in spite of a number of faults, of which the principal were lateness of ships in leaving terminal ports and lack of facilities at the ports along the route. The total freight turnover exceeded the plan by 26 % (237).

1940

Information on the 1940 season is also scanty. There was a large mass of heavy ice in the western part of the Laptev Sea all the summer. In this respect there was a similarity to conditions in 1937, but better organisation prevented any accidents in 1940 (284). The freight plan was over-fulfilled by 10 % (233). Fewer ships were employed than had been the case in 1936, but the turnover of freight was four-fifths as much again (278).

A feature of the Kara operations this year was that double voyages were made for the first time, by two ships: one made two return voyages from Murmansk to Dudinka; and the other did even better, in that its first return voyage was from the west to Tiksi and its second to Dudinka (372).

Through voyages were made, but the exact number is not known. One, from west to east, was made by the German raider *Komet*. An agreement was reached with the Germans that this ship should be passed along the route secretly and as quickly as possible. It is interesting to note that with ice-breaker aid provided at all difficult points it was possible for the ship to sail from Novaya Zemlya to Bering Strait in the record time of 21½ days (of which steaming time was only about 14 days) (9, 48).

Comment (231) by the Head of Glavsevmorput' on the season's activities shows that the results were considered satisfactory, but that "normal commercial exploitation" was still a long way off. Shipping movements, he said, were still too reminiscent of expeditions and not sufficiently like those of a normally working sea route, the objective laid down in the third five-year plan. The next step was declared to be the making of double journeys in one season: ships were to make two return voyages each season from Murmansk to Tiksi, for instance. It was considered that this could be done provided ships could be got out of their home ports on time—lateness in starting was again a fault of the 1940 season—and provided facilities in the Arctic ports permitted a quick turn-round. Such double voyages were planned for 1941.

WARTIME (1941–45)

Information from Soviet sources on wartime activity along the Northern Sea Route is very scarce.

It seems likely that the confusion of the early weeks of the war caused radical modification to the plans for the 1941 season, especially since part of the icebreaker fleet of Glavsevmorput' was taken over by the navy (30). Probably voyages made in 1941 were restricted to essential relief work. In the four following seasons however a large number of successful voyages were apparently made. The Head of Glavsevmorput' claims that the tonnage carried annually increased by 80 % during the period 1940–45 (248). Icebreakers could be spared to escort the vessels since the higher priority task of keeping Arkhangel'sk, Vladivostok and ports in the Okhotsk Sea open to allied supplies as long as possible did not call for the services of icebreakers until after the end of the navigation season on the Northern Sea Route.

The wartime operations were apparently conducted with very small loss. A tug and two barges were attacked by a German naval vessel evidently in the Barents Sea, and were sunk (31). A hydrographic ship, the *Akademik Shokal'skiy*, was torpedoed by a U-boat in the Kara Sea off the north-east coast of Novaya Zemlya in July 1943 (24). The German battleship *Admiral Scheer* entered the Kara Sea in August 1942 and bombarded Ostrov Diksona (125). These are the only losses mentioned by Soviet writers. There is a report of unknown reliability—possibly a German wartime press release—to the effect that the *Admiral Scheer* also sank the icebreaker *Sibiryakov*.

Confirmation that there was considerable activity during the years 1942–45 is forthcoming from American sources.[1] One of the channels by which lend-lease goods reached the U.S.S.R. from the western hemisphere was by way of the Northern Sea Route. Less than $2\frac{1}{2}\%$ of the total volume of lend-lease goods to the Soviet Union went by this route, but nevertheless the tonnage carried during the four seasons amounted to 452,000 gross long tons. The most frequently used course was from North American west-coast ports through Bering Strait to destinations anywhere along the north Siberian shore. It is only possible to establish the Soviet destinations of these ships from the information contained in their manifests completed before leaving America. This information may be misleading, since it is always possible that ships may have been diverted to other ports, or the cargo may have been trans-shipped, after the voyage had begun. But probably diversions would not be made unless absolutely necessary, since most of the ships were to call at several ports and accordingly had their cargo specially disposed for unloading in a certain order. It seems on the whole likely that all the cargo reached its destination in one way or another, since the Soviet authorities sent a general acknowledgment to the United States Government stating that everything had arrived. It may be mentioned that all the ships were entirely Russian-manned, since this ensured their neutral status *vis-à-vis* Japan and therefore lessened the likelihood of their being attacked in the Pacific.

Between 23 and 34 ships were employed each year. Almost all entered the Northern Sea Route from the Pacific; only two came by the Atlantic. The majority were bound for ports east of Mys Chelyuskina. Those most frequently mentioned are Tiksi, Bukhta Provideniya, Ambarchik at the mouth of the Kolyma, and Pevek in Chaunskaya Guba.[2] Presumably the object in keeping most ships east of Mys Chelyuskina was to lessen the chance of their being unable to return to the Pacific in the same season and of being thus unable, or less easily able, to go back to America for another load the following year. In 1945 however some of the ships returned to Arkhangel'sk and Murmansk rather than to Vladivostok (247). There can be no doubt that remarkable voyages were made by some of these ships coming from America; in particular by the Liberty ships, themselves lend-lease goods, which were bigger than any other freighter used up to that time in Soviet Arctic waters. Unfortunately no details are available.

The import of supplies from America constituted a major operation on the Northern Sea Route during the war; but there must have been other very large movements of freight if the statement that turnover increased by 80%

[1] All information on this subject is obtained from ships' manifests, copies of which are included in the files of the Foreign Economic Administration now held by the Department of State in Washington; and from information supplied by Mr Willis C. Armstrong, late of the Foreign Economic Administration of the U.S. Government.

[2] A table giving further information on the movements of lend-lease ships on this route is given in Appendix V.

during the war is true. What these movements may have been is not known. But it may be presumed that the Kara Sea area was serviced by Soviet shipping, as it had been before, and that coastwise traffic between the big rivers was continued. It is also to be presumed that there was some Soviet naval activity in these waters. No details have been published, however; accounts of the wartime doings of the Soviet Northern Fleet make no mention of any event taking place east of the Barents Sea.

POST-WAR PERIOD (1946–49)

Such Soviet publications as have been permitted to leave the Soviet Union since 1946 have contained virtually no information on the working of the Northern Sea Route after the war. Short news items have been released from time to time to the effect that certain icebreakers have been working in this or that area, or that ice-reconnaissance patrols are being continued. The plan for the 1946 season is said (321) to have been successfully fulfilled. In 1949 the press (249) published a report on the successful completion of an operation in which a number of river craft with barges in tow sailed, under icebreaker escort, from Arkhangel'sk to the Ob' and Yenisey where they reinforced the existing river fleets. From this sort of information it is only possible to deduce that activities still go on; their scale remains unknown. A fact which may have some significance is that only two or three British and Norwegian ships have been employed on the Kara operations since the war;[1] whereas before the war, when the Kara operations were conducted on a scale which dwarfed all other activities on the Northern Sea Route, the majority of the ships used were British or Norwegian. But it is known that the Soviet merchant fleet has been greatly enlarged in recent years, so it is conceivable that operations could be continued on the same scale without the use of foreign ships.

(ii) *River shipping*

Shipping on the rivers flowing into the seas crossed by the Northern Sea Route plays an important part, as we have seen earlier, in the development of the sea route itself. During this period from 1933 onwards the great increase in sea-going traffic was accompanied by developments in the sphere of river transport. The outline given below is based, as will be patent, on insufficient and sometimes contradictory sources. For this reason the figures given should not be considered as accurate, but rather as approximations providing a reasonably reliable general impression.

Before 1933 the only rivers which were navigated to any considerable extent were the Ob', the Yenisey and the Lena. Navigation on each continued. Since the upper reaches of these rivers are, and have in the past been, normally used for the transport of goods unconnected with the Northern Sea Route, we will consider only the lower reaches of each river and those craft which habitually visit the trans-shipment points at the mouth. The fleets on the lower reaches came under the control of Glavsevmorput' in 1933.

---

[1] In the 1950 season larger-scale operations were planned, calling for some twenty British and Norwegian ships, according to *The Times* of 13 July 1950.

The Ob' fleet in 1933 consisted of four powered vessels and ten barges, with headquarters at Novosibirsk. In 1936 there were only two powered craft and thirteen barges. The freight turnover in 1933 was 32,000 metric tons, in 1936 nearly 59,000 metric tons (151). This was presumably due to the new barges being of larger capacity. During the years that followed the tendency was for the Yenisey to be used by ships of the Kara operations, to the exclusion of the Ob'. It therefore seems reasonable to assume, in the absence of any precise information on the point, that the river fleet on the lower Ob' got no bigger after 1936.

On the Yenisey a fleet of 32 powered vessels and 58 barges, based on Krasnoyarsk, was left by Komseverput' (263). Another source (151) numbers this fleet at 48 powered vessels and 56 barges. By 1936 there were 39 powered vessels and 75 barges. The cargo capacity of the barges was now 48,000 metric tons as against 16,000 metric tons in 1933, and the freight turnover rose from 22,000 to 70,000 (151). This again was due to the introduction of large-capacity barges. At the end of the 1936 season forty vessels of the fleet were caught by the ice after an expedition up the Pyasina, and they had to winter on that river (167). There were therefore difficulties in 1937, but the plan for freight carriage in 1938 was over-fulfilled. In view of the increasing importance of Igarka, it seems likely that the fleet on the lower Yenisey continued to grow; though it must be remembered that Igarka was essentially a timber centre and its raw material could be rafted downstream to it without need for vessels of any sort.

The river fleets on the lower Yenisey and Ob' were taken out of the control of Glavsevmorput' probably in 1940, but those on the Yenisey continued to play an important part as far as the Northern Sea Route was concerned.

There was a considerable fleet on the Lena, built up at the end of the nineteenth century and the beginning of the twentieth; but up to 1930 by far the largest part of it was employed on the upper reaches of the river system since that was the principal route by which essential supplies reached Yakutiya. In 1933 there were said to be four powered vessels and one barge working for Glavsevmorput' on the lower river (151). The proportion sounds unlikely, but some of the powered vessels were probably motor-schooners which could not be used for towing work. This fleet was admitted to be quite unable to handle the goods which began to arrive at Tiksi in 1933. It was therefore quickly augmented both with powered craft, the largest of which were sent to the river by sea from the west, and with barges which were probably transferred to the lower river from fleets on the upper river. As a result there were in 1936 nineteen powered craft and 58 barges, and there was a rise in freight turnover from 4800 to 30,000 metric tons (151). There were setbacks also on this river. In 1935 the fleet was frozen in for the winter before it could get back to its base at Yakutsk; and in 1938 the freight plan was under-fulfilled. During the war lend-lease goods destined for places in Yakutiya were sent to Tiksi, so it must be assumed that the river fleet continued to function.

The Kolyma had no river fleet of any size until after 1932. Ships had brought cargo to the mouth of the river intermittently since 1911, but the volume was not sufficient to stimulate formation of an organised river fleet; and there was no river fleet in any other part of the Kolyma basin, for unlike the Ob', Yenisey and Lena, the upper reaches of the Kolyma did not link up with centres of agriculture or industry. In 1932 two tugs and some barges were brought to the river, and these formed the nucleus of the future fleet. It was taken over and increased in size by Dal'stroy, a mining and general development organisation which came into being in 1932 in order to work the rich goldfields recently found on the upper Kolyma. The graph of tonnage carried curves steadily and steeply upwards until in 1937 27 times as much was carried as in 1932. In 1939 Glavsevmorput' took over the running of the fleet from Dal'stroy, and established headquarters at Zyryanka. Over 1800 kilometres of the river were now navigable (127). During the war lend-lease goods reached the mouth of the Kolyma for onward transmission, so one must assume, as in the case of the Lena, that a river fleet was there to deal with them.

The expansion of northern economy demanded the use of more rivers for transport purposes. Several hitherto unexploited rivers were accordingly brought into service.

The Pyasina, which drains the south-eastern part of Taymyr, provides a waterway to a point close to the extensive mineral deposits of the Noril'sk area, and is also a possible link with the not easily accessible Volochanka region. Further, its tributary the Dudypta passes close to a tributary of the Khatanga, and natives had long used this route as a water-road across the base of Taymyr. This fact led to a plan being put forward in 1933 for a south Taymyr waterway. The object was to provide shipping on the Northern Sea Route with an inland waterway which would allow ships to by-pass Mys Chelyuskina, the most northerly and often the most ice-encumbered part of the route. Much would have to be done, of course—canal and lock construction, dredging and widening. The idea was keenly debated for years, but in 1939 things had still got no further than the reconnaissance stage (267). However, the project provided an additional impetus to explore and improve the navigability of the Pyasina. From 1932, when the first hydrological party visited the river, ships of the Yenisey fleet made annual expeditions up the Pyasina (167). The completion however of a narrow-gauge railway between Noril'sk and Dudinka on the Yenisey in 1937 or 1938 relieved the very much longer river route of much traffic. It was the intention in 1939 to continue the railway a further twelve kilometres to the river quay on the Pyasina system, so that in future goods for the Volochanka region might also be sent via Dudinka instead of up the Pyasina from its mouth. The Pyasina thus had no river fleet of its own, and the indications are that its usefulness will have waned unless the south Taymyr waterway has finally become a reality.

The Yana, the next large river east of the Lena, had been navigated up to Verkhoyansk by a hydrological expedition in 1927–28, and had been visited by a tug and a barge from the Lena in 1932. Carriage of freight on the Yana started in 1936 when a tug and three barges arrived on the river (115). More vessels reached the river later. By 1940 the regular routine was established of bringing goods downstream, meeting the sea-going ships which since 1939 called at the mouth, and taking incoming freight back upstream (207). The plan for 1940 was over-fulfilled. Between 1942 and 1945 an average of four ships a year were due to call at the Yana with lend-lease goods.

The Indigirka, east of the Yana, was first visited by a hydrological reconnaissance party in 1930–31, when a map of the middle and lower reaches was made. Another expedition continued the work on the upper reaches in 1931. It was by then established that the river was suitable for navigation, but no use was made of this knowledge for several years. In 1935 a Glavsevmorput' party came to the river to see about construction of quays at various points, and in 1936 the first river craft arrived from the Lena (33). In 1937 Dal'stroy, whose area of operations now stretched to the upper Indigirka, started to use the river for the transport of its stores. The 1938 season was not a success because of bad co-ordination between the river craft and the sea-going ships (126). By 1940 there was a fleet of nine powered vessels and 28 barges. The river was navigable as far as the Moma, a distance of some 1100 kilometres from the mouth (128). The fleet was administered by Glavsevmorput' jointly with the Kolyma fleet (127). About two lend-lease ships were scheduled to call at the mouth of the river each year of the scheme's operation.

West of the Lena work was done on the Khatanga, the Anabar and the Olenek. Hydrological expeditions had worked on the Khatanga from 1934 to 1936 (117), but navigation did not start until 1938, when the first river steamer arrived (167). The extent of navigation is not known. During the war some of the lend-lease goods sent to Nordvik and Kozhevnikovo at the mouth of the Khatanga were addressed to the Khatanga river fleet, which was presumably therefore in existence then. The Anabar was mapped in 1935–36 by a hydrological party, and in 1937 a hydrographic detachment worked on the lower river. In the same year the first tug arrived and from 1938 onwards a freight service, probably only capable of carrying a maximum of 750 metric tons at a time, ran between the mouth and Saskylakh (265). Over the war period, a total of eight lend-lease ships were due to call at the mouth of the Anabar. There is very little information about the Olenek, but a tug came to the mouth of the river in 1937, a lend-lease ship was due to call there in 1942 and again in 1943.

The river transport organisations under Glavsevmorput' do not as a whole appear to have been so successful in forging ahead as the sea transport departments. River transport personnel were frequently rebuked (5,19,126) for their failure to fulfil the plan and for the exceptionally large number of wrecks suffered by their fleets. The causes were variously alleged to be lack of discipline, corruption, lack of skill, bad administration. Some at least of the charges were certainly substantiated. But they are not the only cause for under-fulfilment of plans. For instance, there is no doubt that initially, until 1936 or perhaps later, not enough new powered vessels were sent out to the rivers (7). Nevertheless, the fact that plans were frequently under-fulfilled does not mean that an advance was not made, but that the pace was slower than was intended. It is fair to say that navigation was, in spite of shortcomings, successfully maintained and in the aggregate increased on the three principal rivers; it was built up from a very modest level on the Kolyma; and it was introduced on six northern rivers—the Pyasina, Khatanga, Anabar, Olenek, Yana and Indigirka. In thus rendering new river basins accessible to the Northern Sea Route, an enormous area of land was opened up for possible economic development.

3. THE CHIEF ADMINISTRATION OF THE NORTHERN SEA ROUTE [GLAVSEVMORPUT']

For a proper comprehension of the way the Soviet Government set about the task of developing the Northern Sea Route it is necessary to study the structure and evolution of Glavsevmorput'. The structure of an organisation within the Soviet system is especially important, since under that system initiative comes from the top. An organisation is set up in order to shape events rather than co-ordinate work already being done. Glavsevmorput' was called into being in order to achieve certain ends. By virtue of its constitution and its position in a socialist economic system it is at once a Government department and a business monopoly, and it has therefore very extensive powers.

Glavsevmorput' did not, properly speaking, evolve out of any earlier organisation, though there were two bodies which worked in related spheres. One was Komseverput', of which we have already spoken (see p. 18). Another was the Government Arctic Commission [Pravitel'stvennaya Arkticheskaya Komissiya], a body set up in 1928 in order to formulate and put into effect a five-year plan for Arctic development, to form part of the first five-year plan of 1928–32 (230, 349); but the means at its disposal forced it to keep within very modest bounds. Both these organisations did something to co-ordinate work in the Arctic. But nevertheless in 1932 separate authorities were responsible for shipping through the Kara and Laptev Seas, for shipping through the East Siberian and Chukchi Seas, for the network of polar meteorological stations, and for hydrographic and other expeditionary work. Further, the job of carrying supplies to northern settlements was divided among seven organisations, some of them with overlapping functions. If the Government's plan for a big programme was to succeed, better co-ordination was clearly necessary. It would presumably have been possible to enlarge the structure and extend the scope of Komseverput'. In fact, Komseverput' had been expanding for some years before 1932; but it did not at this time make a very favourable impression on the Government. The new department of Glavsevmorput' was therefore established in Moscow by a decree of the Council of People's Commissars of 17 December 1932. The status of the department was that of a Chief Administration [Glavnoye Upravleniye], which was an independent body directly responsible to the Council of People's Commissars but ranking lower than a People's Commissariat [Naroknyy Komissariat], or Ministry [Ministerstvo], as it later became. By the decree (312, 313) Glavsevmorput' was given the task of "conclusively developing the Northern Sea Route from the White Sea to Bering Strait, of equipping this route, keeping it in good order and securing the safety of shipping along it". All meteorological and radio stations on the shore and islands

of the Arctic Ocean were to be transferred to the control of Glavsevmorput'.
O. Yu. Shmidt was appointed Head.

The decree of 17 December 1932 was a preliminary which went little further
than mention the object of the new department. During the next 3½ years four
more decrees were issued, laying down in very considerable detail the constitution and scope of Glavsevmorput'. By the first of these, dated 11 March
1933 (314), all activities and installations of Komseverput' were handed over to
Glavsevmorput'. A great many faults had been found in the way Komseverput' had discharged its duties: in particular, no accounts were being kept, and
the programme of work to be done was far from completion. Komseverput'
had been responsible, it will be remembered, for a number of industrial
undertakings in the Ob', Yenisey and Lena areas; and in taking over these
Glavsevmorput' at once acquired both commercial interests and responsibility
for affairs that were only indirectly related to the Northern Sea Route.

Further powers were conferred on Glavsevmorput' by a decree of 20 July
1934 (226). It was here made clear that development of the sea route remained
the central task. Ten icebreakers were transferred from the People's Commissariat for Water Transport to Glavsevmorput', which now became the central
authority for all icebreaking activity in the U.S.S.R. The decree outlined what
was expected of Glavsevmorput' in the way of organising river transport and
a hydrographic service. But in addition to these matters concerned with
transport, the area of activity of Glavsevmorput' was increased to include, in
the European part of the Soviet Union, all islands and seas in the Arctic,[1] and
in the Asiatic part, the whole area north of lat. 62° N., the latitude of Yakutsk.
This meant taking over more already established economic enterprises, including Arktikugol' (the coal-mining trust in Spitsbergen), the coal- and
metal-mining area at Noril'sk, the coal-mines at Sangar-Khaya on the Lena,
and certain agricultural, reindeer-farming and fishery undertakings. The
responsibility for geological exploration of this vast area, and for developing
the local production of food within it, was also placed on Glavsevmorput'. The
necessity for training experts to do these newly acquired jobs was emphasised.
The administration was also equipped, as were all large undertakings in the
U.S.S.R., with a political branch which organised a system of communist party
organs extending into every sphere of activity, and was responsible for morale
and the maintenance of communist principles. The tasks of Glavsevmorput'
thus now comprised mastery of the Northern Sea Route and exploration and
exploitation of natural resources over a vast area.

An outline of the structure of Glavsevmorput' was approved by a decree of
the Council of People's Commissars of 28 January 1935 (316). This structure was
designed to take into account the extension of scope decreed in the previous
year. The various industrial enterprises taken over in 1933 and 1934 were
grouped regionally and placed under the control of six local offices known
as District Administrations [Territorial'nyye Upravleniya]. Each district
administration was to function as a self-supporting economic enterprise.
The need for scientific guidance in the performance of the new tasks was

[1] This evidently means inside the Arctic Circle.

recognised in the incorporation of several research institutions, notably the Arctic Institute. It is not appropriate to examine the rest of this decree in detail since it describes only a transitional phase. The expansion of Glavsevmorput' was not yet finished. There was added to its duties in 1936 promotion of the economic and cultural development of the native population of the far north. A further statute, the most detailed yet to be issued, laid down the functions and structure of Glavsevmorput' in the light of all its various duties. This statute was confirmed by a decree of 22 June 1936 (317). Under the terms of the statute, Glavsevmorput' consisted of the chief administration itself, and

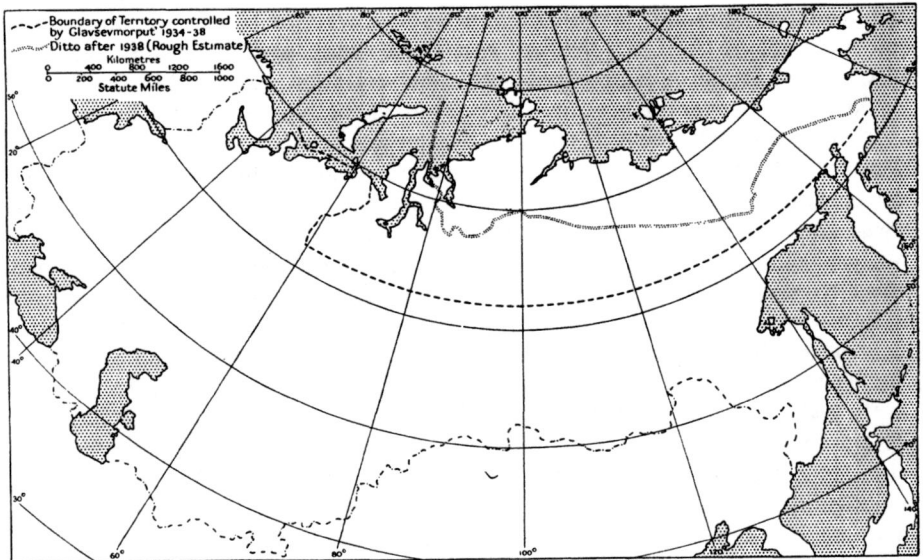

Map 4.   The territory of Glavsevmorput'

a number of subsidiary administrations [upravleniya] and departments [otdely]. A department was smaller than an administration. The administrations were for Political Affairs, Maritime and River Transport, Polar Aviation, Polar Stations (sometimes called the Polar Administration), Hydrography, and Mining and Geology; the departments were for Agricultural Economy, Promotion of Culture among the Natives, Fur Industry, Planning and Economy, Finance and Book-keeping, and Mobilisation of Internal Resources. There was also an office management section. Directly attached to the Head of Glavsevmorput' were a personnel office and the Interdepartmental Bureau of Ice Forecasting. The scientific institutes under Glavsevmorput' control now numbered three: the Arctic Institute, the Institute of Economics of the North [Institut Ekonomiki Severa], and the Hydrographic Institute [Gidrograficheskiy Institut], which was a training school for hydrographers. Shortly afterwards the Institute of the Peoples of the North [Institut Narodov Severa] came into the Glavsevmorput' system. The following organisations also either came or remained within the jurisdiction of Glavsevmorput': an aviation training school, Arktikugol', fluorspar mines at Amderma, Nordvikstroy (a trust investigating mineral deposits at Nordvik), and Arktiksnab (the office for obtaining supplies for ships, expeditions and stations maintained in the north). The district administrations were retained. These now numbered seven and covered the administrative districts of Arkhangel'sk, Omsk, Krasnoyarsk, Yakutsk, the Far East (headquarters at Vladivostok), Leningrad and Murmansk. Each district administration continued to manage industrial concerns in its area, with the exception of those mentioned above, which were under direct control of the chief administration. District administrations also controlled, in their own area, ports, sea- and river-craft, polar stations, hydrography, agriculture, relations with natives, administration, and even certain sea shipping routes. There was in fact a large measure of decentralisation. The corresponding administrations at Moscow, which were ultimately responsible for work in all these fields, acted as policy makers and exercised general supervision over the work of their counterparts in the district administrations. Economic enterprises, such as fur and fishery stations and management of river fleets and ports, were to remain self-supporting units. There was thus a distinction, within each district administration and within the central office, between departments carried on the state budget and financially self-supporting departments.

Glavsevmorput' had now reached the peak of its development as far as power and influence were concerned. Its writ ran over nearly 5,750,000 sq. km. of territory. Its Deputy Head, N. M. Yanson, described it as a "remarkable and many-sided People's Commissariat for northern transport, economy and culture" (392)—a long step from the limited sea transport department envisaged by the decree of 1932. It had grown from nothing in the space of four years. Clearly the Government had great confidence in the organisation and its leaders. This confidence was apparently justified, for the navigation seasons of 1933 to 1936 were all successful. The only major disaster was the loss of the *Chelyuskin* in 1934, but that was turned into a triumph by the spectacular efforts of the airmen who rescued the survivors. But it would be wrong to suppose that everything in this great organisation was running smoothly. There was plenty of inefficiency and stupidity in administration, as one might well expect in a mushroom growth of such complexity. As a result of application of the communist precept of the necessity for criticism and self-criticism, the pages of the journal *Sovetskaya Arktika* teem with examples. In particular the district administrations were criticised. They were accused of inefficiency and corrupt practices, and their accounts often showed large losses(246). As we have already seen, there were shortcomings in the river fleets, which came under the district administrations. Yet it is clear that many officials of Glavsevmorput' were blind to their own faults because of the apparent success of the shipping seasons. The success of the 1936 season led many to feel that the Northern Sea Route had now been mastered and that the main task of Glavsevmorput' had been completed. This combination of self-complacency with inefficiency was clearly unhealthy, and was proclaimed to be so by

Shmidt (290). But as might be expected no drastic reforms were carried out while things were still going well.

In 1937 things went badly: 26 ships were forced to winter at sea; freight and construction plans of all sorts were hopelessly under-fulfilled. Here was a major disaster. Whatever its causes, a thorough investigation of the whole structure of Glavsevmorput' was to be expected. This was done, but not immediately. The Government's first reaction took a political form. During 1937 and 1938 there was taking place in Moscow and elsewhere a series of trials for treason of leading communist figures. Treason was suspected everywhere. If things had gone wrong in the Arctic, the cause, it was thought, could not possibly be anything but treason—deliberate wrecking by hostile elements in the population. The disaster of 1937 was discussed in the Council of People's Commissars in March 1938. The causes were found to be "bad organisation in the work of Glavsevmorput', the existence of complacency and conceit, and also a completely unsatisfactory state of affairs in the choice of workers for Glavsevmorput', all of which created a suitable setting for the criminal anti-soviet activity of wreckers in a number of Glavsevmorput' organs" (225). Reports on the events of 1937 and on plans for 1938 were called for from the Head of Glavsevmorput', in order that the council might assure itself that the same mistakes were not going to be made again; and a purge of "doubtful elements" in Glavsevmorput' was ordered (225). "Doubtful elements" proved to be, among many others, the heads of administrations which had been shown up particularly badly. Many had in fact already been purged in 1937. E. F. Krastin, Head of the Administration of Maritime and River Transport, and also one of the Deputy Heads of Glavsevmorput', was branded as an enemy of the people and dismissed (350), in spite of the fact that he had not long previously received the Order of Lenin in recognition of the successful season of 1936. S. A. Bergavinov, (261) Head of the Political Administration, and P. V. Orlovskiy (19), Head of the Hydrographic Administration, suffered the same fate. Several district administration heads were also dismissed (19). Besides these leading officials a large number of rank-and-file workers went too. The conduct of the purge as a whole evidently met with official approval, for the new Head of the Political Administration, L. Yu. Belakhov, was promoted and awarded the Order of Lenin in 1940 for his good work in the previous two years (347).

This large-scale political purge probably had essentially the same effect as a normal inquiry into inefficiency would have had. The difference was the degree of savagery with which it was carried out. There is no doubt that bad organisation had much to do with the catastrophe of 1937. But it is disconcerting to western eyes to read that a man who has been unpunctual in getting his flotilla of river craft to its destination is therefore a Trotskyite-Bukharinite spy and bandit. The lengths to which the purgers went in order to find political reasons for inefficiency were sometimes extraordinary. For instance, the fact that polar station radio operators who were trained at Moscow received different rates of pay from those trained at Leningrad was attributed to a subtle and deliberate wrecking policy, designed to cause jealousy and therefore bad work (348). Injustice was no doubt frequently done, but Glavsevmorput' probably emerged with fewer inefficient officials. Reorganisation of the structure was now undertaken.

The Council of People's Commissars again discussed Glavsevmorput' in August 1938. The main cause of the disaster, now that the spy scare was dying down, was seen thus: "Having enlarged its functions and divided its attention among the various branches of its economy, Glavsevmorput' did not secure the economic management of these branches and in particular paid far too little attention to its principal task—the mastery of the Northern Sea Route (242)." There can be little doubt that this was a correct analysis. The action taken as a result was to restrict the activities of Glavsevmorput' to matters relating directly to the Northern Sea Route, with the emphasis on the much less well studied eastern sector. This could now be done because "the formation in recent years of a number of new provinces in the northern regions, and the resulting strengthening of the Soviet machine there, permits the organisation of an economic and cultural service for the population of these regions, and allows Glavsevmorput' to be released from these tasks" (242). The district administrations were abolished, and their responsibilities for sea and river transport, local trade and hydrography were assumed by the Glavsevmorput' administrations in Moscow. General responsibility for cultural and educational work and for the development of economic resources north of the 62nd parallel passed to the appropriate commissariats of the republics concerned (242). Most of the fisheries and airlines operated by Glavsevmorput' were transferred to other authorities. Glavsevmorput' retained control only over the coastal areas, including Taymyr, northern Yakutiya and Chukotka (171) (see Map 4 on p. 56); and over the coal-mines at Sangar-Khaya on the Lena, and a shipyard at Peleduy on the same river. Outside these areas Glavsevmorput' was only permitted to have representatives in Igarka, Yakutsk and Anadyr' (242). It was realised by all that a great deal more work remained to be done on the sea route before anyone could again say that it was conquered. Thus it was that the task set for Glavsevmorput' in the third five-year plan sounded so similar to the task set in 1932: V. M. Molotov said in his speech before the Eighteenth Party Congress in March 1939 that one of the objectives in the sphere of transport was "by the end of the third five-year plan [1942] to turn the Northern Sea Route into a normally working waterway, securing a regular link with the Far East" (342).

Apart from this major change, the constitution of Glavsevmorput' was constantly being modified to meet new needs. The first minor modifications were introduced while the bigger policy changes were still under consideration. A decree of 25 June 1938 (306) authorised the formation of certain new departments. An Administration of Construction was formed to deal with the extensive building programmes undertaken at ports and elsewhere. Responsibility for the river fleets was taken from the Administration of Maritime and River Transport and vested in a new Department of River Transport—a sign that greater attention was to be paid to the river fleets, probably in view of both their recent inefficiency and the present pioneering of more rivers in north-eastern Siberia. Among the innovations were a Department for Training Cadres, and a Control and Inspection Group; both were probably intended to act as safeguards against any more "doubtful elements". The Administration of Maritime Transport was abolished by a decree of 1 June 1940(308), and responsibility for organising shipping was concentrated in a specially formed staff directly under the control of the Head of Glavsevmorput'. By the same decree Nordvikstroy was dissolved because the attempt to mine minerals at Nordvik was not meeting with success. The Mining and Geological Administration took over responsibility for geological reconnaissance here as elsewhere in the diminished area controlled by Glavsevmorput'.

The next decree, of 25 January 1941,[1] is worth studying more fully since it is the most recent statement of the structure of Glavsevmorput', and also the first complete picture since the statute of 1936 which it replaced. The extent of the changes of 1938 is clearly visible; so also is the shift of emphasis to the eastern sector of the route. There are no great changes in the administrations, which are now as follows: Political Affairs, Arctic Fleet and Ports, River Fleet, Polar Aviation, Hydrography, Polar Stations, Mining and Geology, Capital Construction, and Arctic Supply (Arktiksnab). The Administration of the Arctic Fleet and Ports evidently fulfils at least some of the functions of the old Administration of Maritime Transport; the decree does not make it clear to what extent the arrangement of subordinating shipping directly to the Head of Glavsevmorput' still stands. Both Arktiksnab and the Administration of Capital Construction were placed on a cost-accounting basis. Of the departments, those of Agricultural Economy, Promotion of Culture among the Natives, Fur Industry, Trade and Mobilisation of Internal Resources no longer exist, having presumably been dropped in 1938. The remaining departments dating from the 1936 statute and subsequent decrees were retained, though sometimes with some telescoping of duties: the Planning and Financial Department, the Central Book-keeping Department, the Departments of Teaching Institutions and Preparation of Cadres, and of Work and Pay; the Control and Inspection Group, and the administrative section. To these were added a Department of Leading Cadres (evidently a personnel selection bureau for the higher posts); a War Department, and an Inspectorate of Security and Anti-Aircraft Defence, necessitated by the war in Europe; and some minor offices.

An appendix to the decree lists all undertakings and organisations responsible to Glavsevmorput'. The list gives a clear idea which branches had been expanding and which contracting. Sea transport is placed in the hands of two newly formed Arctic sea shipping companies [arkticheskiye parokhodstva] based at Arkhangel'sk and Vladivostok, with another office and the main repair yard at Murmansk. The three ports of Dikson, Tiksi and Provideniya were by now sufficiently developed to house administrative offices working on problems of sea transport, and each of these ports is equipped with a building unit. River shipping controlled by Glavsevmorput' was now restricted to rivers east of Taymyr. In support of the river fleet, and to a certain extent of

[1] A full translation is at Appendix VI.

the sea fleet also, there were two yards for building wooden ships and lighters at Arkhangel'sk and Peleduy on the upper Lena, and a yard for building metal ships at Kachuga, also on the upper Lena. The hydrographic service was considerably enlarged. Mining centres now numbered only three: coal at Sangar-Khaya, and oil reconnaissance parties at Nordvik and Ust'-Port. Arktikugol' had evidently been transferred to other authorities (the war did not compel the closing of the Spitsbergen mines until six months after the date of this decree). Aircraft were organised in four groups, based at Moscow, Igarka, the Lena and Chukotka, with the main repair centre at Krasnoyarsk. The number of scientific and educational institutions affiliated to Glavsevmorput' increased. The Arctic Institute remained the principal scientific organ. The Institute of Economics of the North was dissolved soon after 1936, but its place was taken by a special economic branch of the Arctic Institute. The Institute of the Peoples of the North became irrelevant to Glavsevmorput' after 1938 and was transferred out of its control. The Hydrographic Institute remained. For the training of scientific and technical staff there were formed a Hydrographic Technical School at Leningrad, a Marine Technical School at Murmansk and a Technical School of Meteorology and Communications at Moscow.

Since 1941 no information on any alterations in the policy or structure of Glavsevmorput' has been published. Since such information has not been withheld in respect of similar organisations, it may be reasonable to assume that there has been no large change, although the possibility that the structure of Glavsevmorput' is now on the secret list cannot be ruled out. The task facing Glavsevmorput' after the war was indeed very much the same as it had been just before the war. The third five-year plan had been interrupted before its objects had been achieved. Therefore the fourth five-year plan of 1946–50, of which the principal object was making good wartime losses, prescribed that the Northern Sea Route should become a normally working sea lane by 1950 (143).

It must not be forgotten that the Soviet communist approach to all problems has been an essential feature of Glavsevmorput' during the whole of its existence. The methods employed have been those common to all large undertakings in the U.S.S.R. First of all, the activities of Glavsevmorput' were included in the framework of the five-year plans. These plans served two ends. First, they co-ordinated the needs and output of each department with the rest of the country's economy. Secondly, they were made to act as incentives to work by being broken down into narrowly defined short-term objectives which were placed before subdepartments, groups of workers and sometimes individuals; even the staff of scientific institutes had to fulfil detailed plans. The second use was in this instance probably more important than the first. The Stakhanovite method of work—of which the most important feature is payment by piece-work—was introduced into all branches of Glavsevmorput' after 1935, at the same time as it was being introduced into industrial concerns all over the country (22). "Socialist emulation"—in which one group of workers challenges another to exceed its output—is constantly encouraged between units of similar type, from icebreaker crews to winterers at polar stations (208).

Komsomol, the communist youth league, is very active in Glavsevmorput';
at one time the *Krasin* was manned entirely by Komsomol members (319). In
1938 each of the main administrations was given the Soviet decoration of the
Red Banner, to be awarded annually to the most efficient unit within that
administration, on the same principle as the challenge cup. In addition
individuals who had distinguished themselves were awarded special badges,
sometimes coupled with money prizes (245). It was the duty of the Political
Administration to put across these methods of getting people to work harder.
For this purpose a considerable propaganda machine was developed, using
the radio and a number of locally produced news-sheets to reach the many
outlying stations where Glavsevmorput' employees worked. The Political
Administration was always listed first among the administrations and its work
was clearly considered of fundamental importance.

Something should be said of the workers themselves. Glavsevmorput'
employed a large number. The total in 1936 was given as 32,352 (394). This
number probably increased up to 1938, and may then have grown smaller
after the curtailment of many Glavsevmorput' functions. Since the early days
of Glavsevmorput' there have been special privileges granted by decree (310, 315)
to workers in the far north—extra pay, extra leave, better pensions and many
other advantages. These, together with the glamour which attaches to work
in the Arctic, have been a sufficient incentive to attract skilled workers. There
is evidence (34, 192, 305) to show that some of the large industrial enterprises in
the north, for instance at Noril'sk, in the Pechora coalfields, and the Dal'stroy
organisation on the upper Kolyma, are manned by convict labour. Of these
undertakings, only Noril'sk has been managed by Glavsevmorput', and that
arrangement stopped in 1938. It is probable therefore that Glavsevmorput'
employs little convict labour; in any case such labour would not be suitable
for scientific or transport work. Of the free labour employed it is possible that
some may have been directed to their place of work.

The personalities who held the post of Head of Glavsevmorput' deserve
attention. O. Yu. Shmidt, the first, was a remarkable and many-sided man.
He was a university teacher of mathematics before the Revolution. After it
he held a number of senior civil service posts in various commissariats, and
was then in turn Head of the State Publishing House [Gosizdat], Deputy Head
of the Central Statistical Administration [Tsentral'noye Statisticheskoye
Upravleniye), and a member of the Presidium of the State Planning Commission [Gosplan]. During this period he continued to be a professor of mathematics at Moscow and published a number of mathematical works. In 1929
he made his first Arctic expedition, to Zemlya Frantsa-Iosifa [Franz Josef
Land]. In 1930 he became Director of the Arctic Institute and in 1932 was
elected a member of the Academy of Sciences [Akademiya Nauk], a body even
more select than the British Royal Society. In the same year he became Head
of Glavsevmorput' (87). Shmidt was clearly an outstanding man from the
Government's point of view; he was a proved and loyal administrator with
a scientific background of no little distinction. It is possible to argue that the
fact of a senior civil servant and reliable party man of his standing going off

PLATE III

OTTO YUL'YEVICH SHMIDT (born 1891),
Head of Glavsevmorput' from its creation until 1939.

PLATE IV

IVAN DMITRIYEVICH PAPANIN (born 1894), Head of Glavsevmorput'
from 1939 to 1946. Photograph taken at the North Pole in 1937.

to the Arctic in 1929 is so strange that it may indicate that the Government was even at that time thinking about Glavsevmorput' and was training a suitable head. After his appointment to Glavsevmorput' Shmidt continued to take a very active part in Arctic exploration, and was generally aboard one of the icebreakers all through each navigation season. He was many times decorated for his work. His tenure of office was eventful and included the early rapid expansion and the upheaval of 1937–38. He was relieved of his post in March 1939 "in connection with his transference to other work" (307). He was certainly lucky to avoid the purge, for distinction and ability were not in other cases sufficient to save people. Possibly it was thought inadvisable to call the hero of the *Chelyuskin* and other expeditions an enemy of the people. It is not known what "other work" he was transferred to; but he has continued to contribute to learned journals since 1939, and was still in 1948 editor of the *Great Soviet Encyclopedia* [*Bol'shaya Sovetskaya Entsiklopediya*]—a post he has held since 1983—and of the geographical and geophysical series of the *News* [*Izvestiya*] of the Academy of Sciences.

Shmidt was succeeded by I. D. Papanin, a very different type of man. The son of a sailor, he was a metal worker at Sevastopol' in his youth. When the 1914–18 war broke out he was twenty and joined the Imperial Navy. After the Revolution and civil war—in which he fought energetically for the communists—he held various small jobs. He first went to the Arctic in 1931, and in the years that followed was the leader of several wintering parties at polar stations (87). He became famous as the leader of the North Polar Drift Expedition of 1937–38, on which four men drifted on the sea ice across part of the central polar basin. During the purge period he was on that remarkable expedition, and this not only gained him much publicity, but kept him quite free of the suspicions which seemed to attach to almost anyone then taking an active part in the administration of Glavsevmorput'. Immediately on his return he was made Senior Deputy Head. He was not a scientist, nor had he had a university education. But he had a good knowledge of the Arctic, he had shown his ability to lead men, and above all he was a thoroughly reliable party man. His speeches contain eulogies of the party chiefs that are remarkable for their fulsomeness even in a good party man. His tenure of office as Head of Glavsevmorput' included the whole war period. At first he had to see that the changes decreed in 1938 were put into effect. The shipping seasons of 1939 and 1940 show that he obtained quite good results. Of the war period we know little, but it would seem that he obtained the results required of him. He was given the rank of Kontr-Admiral in the Soviet navy. In June 1946 he retired on account of illness.

Papanin's place was taken by A. A. Afanas'yev, who until then had been Deputy Minister of the Sea Fleet (311). Very little is known about him. Possibly he can be identified with that Afanas'yev who was in 1941 in charge of the Far Eastern Shipping Company at Vladivostok. Certainly the appointment of a shipping expert would be appropriate. As far as is known, Afanas'yev was still in office in 1949.

The foregoing gives some idea of the evolution of Glavsevmorput', at least during the first ten years of its existence. It grew rapidly, became complicated and unwieldy in structure, and tried to do too much. It was then relieved of responsibility for affairs not directly concerned with the sea route. This had a salutary effect, and served to keep the pioneer character of the organisation, which was still a very considerable size. In tracing this outline we have been concerned with some activities of Glavsevmorput' which are not directly relevant to our main theme; but notwithstanding this, the importance of Glavsevmorput' in the history of the Northern Sea Route remains paramount. For the sea route was always the main preoccupation of Glavsevmorput', even when the latter had many other preoccupations. And with one possible exception the sea route was not the concern of any other organisation. The exception is the Soviet navy, which may have been active in the area. There is no published information on this point. The navy was almost certainly interested in the possibility of using the route, and probably assumed control of some activities of Glavsevmorput' during the war—Papanin's naval rank tends to confirm this. But it seems likely that the navy, which before the second world war was very small, would have left to Glavsevmorput' all the work of equipping and maintaining the route. It is reasonable to conclude therefore that Glavsevmorput' must be held solely responsible for the successes and failures in the development of the Northern Sea Route since 1932.

4. EQUIPMENT

(i) Ports

Glavsevmorput' inherited very few ports and few of those were much more than anchorages. The Kara Sea traffic had been handled at Novyy Port or Igarka. Both were located several hundred miles from the open sea, up the estuaries of the rivers they served. The only port which lay close to the principal shipping lane and which could be used as a port of call and coaling station for ships continuing eastwards was Ostrov Diksona. At the eastern end of the route Nizhnekolymsk had been used by small craft bringing cargoes to the Kolyma. There was no generally used anchorage east of here, though various bays along the shore of Chukotka had been used from time to time. There was therefore a great deal to be done by Glavsevmorput'. Two main needs had to be considered in selection of sites for ports: provision of bases and coaling stations, and provision of ports through which the economic opening-up of the mainland was to take place. The two did not always coincide.

The most satisfactory way of showing how Glavsevmorput' dealt with the problem is to list all ports known to have been used and describe what measures were taken to improve them.

KHABAROVO is a fishing settlement on the south shore of Yugorskiy Shar. It was used occasionally as an anchorage by ships passing through the strait, but it has rarely been used since the 1920's. As far as is known, there are no port facilities there. It has been suggested (395), however, that a railway should be built from the coalfields at Vorkuta to Yugorskiy Shar, a distance of about 150 miles. If this were done Khabarovo is the likely terminus, and the port would immediately become very important as a coaling station. But much capital construction work would have to be done before ships of any size could dock here.

AMDERMA, a few miles east of Khabarovo at the Kara Sea entrance to Yugorskiy Shar, was a similar fishing settlement. In 1933 mining of fluorspar was started here, and loading facilities of some sort must have been provided in the years that followed, since the fluorspar could only be got away by sea. It is likely that the undoubted value of these deposits to the state (they are estimated to contain 48% of the total reserve of fluorspar in the Soviet Union (162)) will lead, if it has not already done so, to considerable development of Amderma as a port.

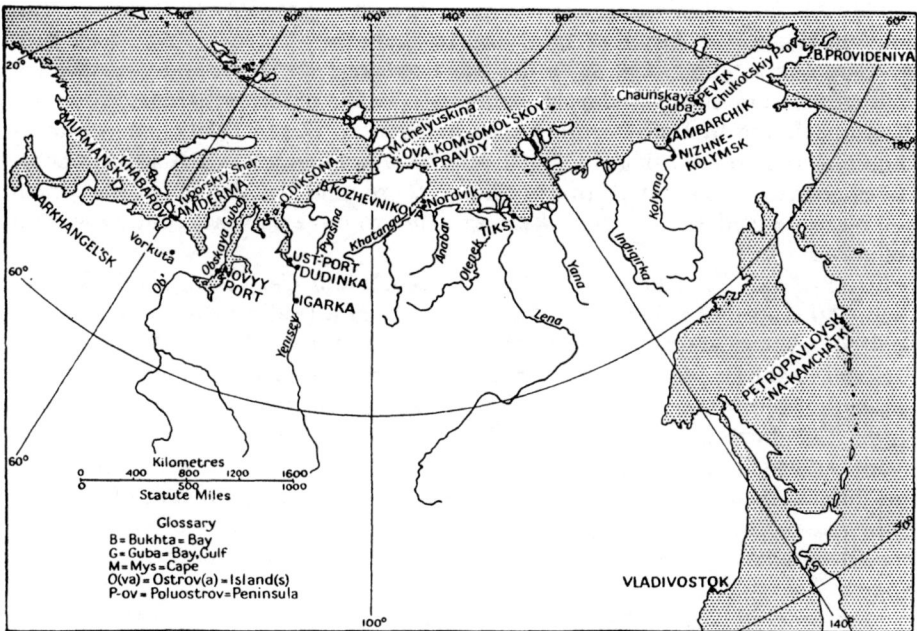

Map 5. Ports on the Northern Sea Route

NOVYY PORT in Obskaya Guba was much used in the 1920's. We have already spoken (see p. 20) of its disadvantages—shallow water and lack of protection from the east. There was never any port equipment there, and under Glavsevmorput' it gradually lost almost all its custom until in 1939 the Kara operations were routed entirely to the Yenisey (179).

The Yenisey can be navigated for nearly 400 sea miles by sea-going vessels drawing up to 22 ft.—that is, with a carrying capacity of about 3000 metric tons. Various places have been used as ports. The most important is IGARKA, of which we have also spoken earlier (see pp. 20–21). Igarka has always been something of a show-piece. Its population rose from 49 in 1927 to 20,000 in 1939 (235). It became the centre of a new and growing timber industry, and was held up as a model of the town planned and built, despite all sorts of natural obstacles, to meet an economic need. Its growth was certainly remarkable, but some of its shortcomings were glossed over. Even at this most important and flourishing port, there were no docking installations until 1935—that is, during the first seven years of its existence. In that year a wooden quay, some of which was temporary, was put up, allowing eight ships to dock at once along the river-front (26). In 1939 there was about 1000 yd. of quay (166), or enough for perhaps ten ships—not a very great increase, when it is remembered that the shortness of the season makes quick turn-round and therefore adequate quay frontage of cardinal importance. On shore, the rapidly growing population caused difficulties in food supply. One of the pioneers and leading officials of the town admitted that there were 3500 cases of scurvy in 1933 (232). Nevertheless Igarka was the largest and probably the best equipped port in the Soviet Arctic certainly up to the war.

DUDINKA, about 136 sea miles downstream from Igarka, is a settlement which was sometimes used as an anchorage by the nineteenth-century traders. It was eclipsed by Ust'-Port and then by Igarka. But in 1939 it came into use again as the port serving the Noril'sk mines, with which it had shortly before been connected by rail. Ships bunkered there, and the size of the settlement increased considerably. Dudinka is said (180) to have a quay standing in 22 ft. of water.

UST'-PORT (also known as Ust'-Yeniseysk and Ust'-Yeniseyskiy Port) is some 60 sea miles below Dudinka. It had been a trading station and port of call for river steamers before the Revolution, and from 1920 to 1928 it was the regular trans-shipment point for the Yenisey (see p. 20). It had a landing stage but the water was too shallow for sea-going vessels to reach it. From 1930 Ust'-Port was the centre of fisheries on the lower Yenisey and a fish cannery was built there (39). In 1939 boring for oil started in the vicinity, but nothing very much had come of this at least up to the end of the war (see p. 87). The size of the settlement grew, but it would seem that only if oil is found in usable quantities will Ust'-Port have a future as anything more than a fishing port.

Presumably any of the other anchorages on the lower Yenisey used by Wiggins, Lied and the rest could still be used. But there is no evidence that any of them have been used, and in fact with Dudinka and Igarka in use there is no need for them.

OSTROV DIKSONA, or Dikson, an island just north of the entrance to the Yenisey estuary, is a good natural harbour and has been used certainly since Nordenskiöld first anchored there in 1875, and possibly before that. A wireless and meteorological station has been established on the island since 1916, and it has been used since 1920 as the base from which the Kara Sea operations were directed. Ships were coaled at Dikson during these expeditions, but probably coaling was carried out from ship to ship in the harbour. In 1934 a coal dump was established ashore (346). In 1936 a wooden quay with a 60-m. frontage was floated down from Igarka and put up on Ostrov Konus, a small islet adjoining Dikson (290). This became the coaling base. Practically nothing else was provided in the way of equipment until 1939. There were bitter complaints in 1937 and 1938 of the lack of loading equipment, of small boats for harbour use, of warehouses and workshops ashore, of adequate quays with sufficient draught (3,6). Early in 1939 workshops were built, and some small boats for the harbour and some mechanical equipment arrived. It is estimated (269) that 28 % of the turnover was handled mechanically in 1939, when 114 ships called. As one would expect, the bulk of the turnover—seven-eighths in 1939—was coal. Bunkering was speeded up from 120 metric tons a day in 1938 to 500 metric tons a day in 1939 (119). In 1940 mechanisation was increased: four coal conveyors came into use. A deep water quay 110 m. long was towed from Igarka to Dikson in August (25), but it was not expected that the quay would be in place and that the port would be fully mechanised before 1942 (8). Dikson was now the principal base for the western sector of the Northern Sea Route, being the centre of the radio network and of the weather and ice forecasting service, and also an air base. Coastwise traffic relayed cargoes from Dikson to the Pyasina and islands in the Kara Sea.

Between the Yenisey and the Khatanga there are no ports. A coaling depot was to be established in 1938 on OSTROVA KOMSOMOL'SKOY PRAVDY (228), a group of islands a short distance south-east of Mys Chelyuskina; so evidently there is at least a suitable anchorage here.

Ships have called at the estuary of the Khatanga since 1933, principally because minerals were suspected in the Nordvik area. The principal anchorages are in BUKHTA KOZHEVNIKOVA and BUKHTA SYNDASKO, adjacent bays on the south shore of the estuary. Bukhta Syndasko is the trans-shipment point for goods destined for the Khatanga river basin, while Bukhta Kozhevnikova is the terminus for goods coming to Nordvik. A quay of unknown draught and some buildings were put up at Bukhta Kozhevnikova in 1939 (160). It is unlikely that the water at the quayside is more than a few feet deep because the shallowness of the bay has often caused slow turn-round of ships— in 1937 it averaged 23 days (4). During the war some thousands of tons of salt were shipped out, but the loading was probably done from lighters.

At the mouths of the rivers Anabar and Olenek there are no ports. Transshipment takes place at any convenient point near the mouth.

The next port east of the Khatanga is TIKSI. The site, in a bay at the southeastern edge of the Lena delta, was selected in 1933 and was first used in that year. In choosing this site Glavsevmorput' wanted to find a place which would serve equally as a trans-shipment point for Lena traffic, and as a coaling station and base for the central part of the Northern Sea Route. The fact that river craft have to cross some miles of open sea to reach Tiksi is of course a disadvantage, but this was unavoidable if sea-going vessels were to reach the port at all. A mole was built in time for the 1934 season but it was not in deep enough water to accommodate anything more than small craft. Later, rails were laid along the mole. In 1937 loading and unloading still had to be done by lighters, and there was no mechanisation at all (3). The average time taken to turn a ship round was nine days (397). In 1939, when the port was visited by twenty sea-going ships and 57 river craft (406), there was little improvement. All loading and unloading was still done by hand; there were not enough barges to ply between ships and shore; bunkering was at the rate of 230 metric tons a day (119). It was only in 1940 that equipment began to arrive. This included a dozen conveyor belts, of which the biggest was 200 m. long and capable of moving 120 metric tons an hour. Cranes, more rails and wagons and other equipment also arrived, but much of it came too late to be used that year. As a result only two ships could be handled at once, and this caused delay; nevertheless the plan was exceeded in the matter of turn-round of ships (120). A quay of sufficient depth to accommodate sea-going ships was to be built in time for the 1941 season, and this was probably done. Meanwhile the shore settlement was increasing in size—the permanent population was 700 in 1939 (374)—and warehouses and workshops were built. The port had to serve more purposes than Dikson, and its turnover in 1939 was nearly equally divided between coal and other goods (119). In 1941 Tiksi was probably beginning to be able to fulfil its functions adequately as the main base of Glavsevmorput' in the area. During the war it was clearly very active, for more lend-lease ships were due to call there than at any other Soviet Arctic port (see Appendix V).

There are no ports at the mouth of the Yana or Indigirka. Transshipment takes place in the roadstead in each case. Things are not very much better at the Kolyma. When navigation to the Kolyma started in 1911 ships generally went up to NIZHNEKOLYMSK. But this could only be done by ships drawing less than 12 ft. When larger ships called at the Kolyma, their cargoes were trans-shipped off AMBARCHIK, along the coast some eighteen sea miles east of the eastern arm of the delta. There has been a mole at Ambarchik since 1933 certainly, but it is only accessible to lighters and shallow-draught vessels. Trans-shipping has to take place from three to ten sea miles offshore in a completely exposed place (43, 161). The importance of Ambarchik increased as the Dal'stroy mining organisation on the upper Kolyma grew larger. Ambarchik became a coaling station in 1938 (37), and coastwise traffic to the Indigirka and Chaunskaya Guba radiated from it. There is a settlement there which includes warehouses. Lend-lease shipping was frequently routed to the port. As a port site Ambarchik has obvious and grave disadvantages; but there appears to be no alternative.

Chaunskaya Guba was listed as a coaling station that was to be used in 1938 (397). Just which part of this very large bay was to be the site, and whether it was in fact used or not, is not known. It is true however that lend-lease ships were directed to PEVEK, on the eastern shore of the bay. Certainly deep water is found close inshore at this point. The need for a port here has probably arisen in connection with tin-mining operations which are thought to be in progress in the vicinity (see p. 106).

Along the coast between Chaunskaya Guba and Bering Strait there are a number of settlements, and small ships call at many of these more or less regularly. There is no pressing need for a port on this stretch, and in any case the anchorages are too shallow and unprotected to be much use.

It became apparent, soon after the creation of Glavsevmorput', that a port and coaling station was badly needed somewhere in the vicinity of Bering

Strait. While there was no port there ships had to go from Petropavlovsk-na-Kamchatke to Tiksi without refuelling—a distance of about 3000 sea miles. This need had been felt before: the *Taymyr* and the *Vaygach* had coaled (from another ship) at BUKHTA PROVIDENIYA, on the south shore of Chukotskiy Poluostrov. It was a long narrow inlet with deep water and steep-to sides. There was no settlement in the bay but it was the best anchorage in the neighbourhood. Glavsevmorput' decided to use it. In 1933 a coal dump was established on a spit, but nothing more was done until 1938. In that year and the next a temporary deep-water quay 30 m. long was built and buildings were put up on shore to house dock workers (121). The first items of mechanical equipment arrived in 1940, in the shape of twelve conveyor belts and narrow-gauge rails, and more houses, including two warehouses, were built (222). A larger and more permanent quay was to be ready for the 1941 season (290). In 1939, 42 sea-going ships called, and this severely strained the port staff. The turnover was two-thirds coal (119). Bukhta Provideniya, like Dikson, is a coaling station and base for various Glavsevmorput' services, but does not itself serve any production centre.

However extenuating may be the circumstances of building ports in these latitudes, it is clear from the list above that up to 1940 the ports and port facilities available on the Northern Sea Route were not impressive. The lack of facilities was by 1939 having serious consequences, for the excessively slow turn-round at most ports was endangering the whole freighting plan. P. P. Shirshov, Deputy Head of Glavsevmorput', wrote in 1939 (285): "At present our ports in no way satisfy the most elementary demands." The third five-year plan envisaged three principal ports and coaling stations—Dikson, Tiksi and Provideniya. As we have seen, these three did begin to receive equipment in 1939 and 1940. As far as can be judged, they remained the most important ports during the war. It is worthy of note that of these three only Tiksi was sited with the intention that it should become an import and export centre. The fact that two of the three principal ports had no such significance tends to show that Glavsevmorput' was at this time more concerned with the immediate objective of securing the efficient working of the sea route than with preparing the way for the economic development which was presumably to be one of the ultimate justifications of the route's existence.

Of the terminal ports of ARKHANGEL'SK, MURMANSK and VLADI-VOSTOK little need be said. None of them was particularly big, but all were well capable of dealing with the Northern Sea Route traffic, which represented only a small proportion of the total number of ships with which they had to deal. All three were enlarged during the period under review, and particularly during the war; but this was related more to the increase in sea-borne freight traffic as a whole than to that of the Northern Sea Route in particular. The importance of PETROPAVLOVSK-NA-KAMCHATKE as a subsidiary base to Vladivostok increased shortly before and during the war. The anchorage here has been used since the eighteenth century. In spite of its great distance from a railhead and from industrial centres, the fact that it is half-way along the route from Vladivostok to Bering Strait gives it obvious advantages.

The problem of repair facilities was more difficult, however, Ships, particularly icebreakers, frequently got damaged in the ice, sometimes seriously. None of the ports between Novaya Zemlya and Bering Strait could undertake any repairs to ships; the workshops at Dikson and Tiksi could only deal with very small jobs. Repairs were generally carried out at Leningrad. Glavsevmorput' badly needed a repair yard of its own, and in 1935 work was started on building one at Murmansk. There were delays and difficulties, and the estimated cost was far exceeded, but building went on and some shops were undertaking work in 1939. The whole yard was to be ready in 1940 (328). It was to contain a dock for *Stalin*-class icebreakers, and was expected to solve most of the repair problems of Glavsevmorput'. Unfortunately there is no information on how well it worked.

(ii) Ships

It is probably true to say that Russia has always possessed a larger fleet of icebreakers than any other country, since their services are required at almost all her ports—in the White Sea, the Baltic, the Black Sea, the Sea of Azov and the Pacific. It is difficult to establish just when and where the first icebreaker was built, since early icebreakers were nothing more than tugs with reinforced bows, and their job was to clear passages through ice inside, or in the immediate vicinity of, harbours. It is generally claimed (153) in Russia that the Russian merchant Britnev was the first to conceive the idea when he modified the *Paylot*, an 85-h.p. steel tug, for icebreaking purposes at Kronshtadt in 1864. It is clear however that such a craft—*City Ice-Boat No. 1*—was built in America in 1837 for use on the Delaware (258); but it is claimed by one Soviet writer (391) that the Russians knew about using small craft for breaking ice even before that. Whatever the truth may be, it is certain that in the last three decades of the nineteenth century a number of small icebreakers were built by Germany, Sweden, Denmark, the United States and Russia. These were used, with varying success, in river estuaries and harbours, principally in the Baltic, the Great Lakes of America, and the Far East. By the turn of the century Russia had about forty small icebreaking craft (240).

The idea of building a much larger and more powerful icebreaker for use in the Arctic occurred to a Russian naval officer, Admiral S. O. Makarov, who was thinking principally of the advantages it would bring to the Kara Sea route. He studied the design of previous icebreaking craft, and also that of Nansen's *Fram*. He had a struggle to get his idea approved, but finally he got permission for Armstrong-Whitworth at Newcastle to build the icebreaker he had been designing. This was launched in 1898 and called the *Yermak*. She was 4955 gross registered tons, with three screws aft and one forward, driven by coal-fired steam reciprocating engines developing 10,000 h.p. Tests at Kronshtadt were successful, but in Arctic waters several defects became clear. For instance the bow screw, which was an American idea designed to help clear away ice fragments and to increase manoeuvrability when going astern, was found to get damaged too easily and so was removed. Another Arctic voyage was made on which the ship had a rather difficult time in the northern part of the Barents Sea. Makarov had maintained that his icebreaker would be able to steam straight to the North Pole. Thus there was a certain anti-climax when it was found that some types of ice after all could defeat even the *Yermak*. This, together with the fact that the ship's hull suffered some damage from the ice, made an unfavourable impression on the Russian naval authorities. Makarov could get no more interest shown in his ship, and he himself was killed in the Russo-Japanese war (152). In 1905, it is true, the *Yermak* was ordered to escort the big Kara Sea expedition of that year, but she went aground off Vaygach and was thus unable to show her worth. The idea and the ship were not forgotten, however. The *Yermak* was the prototype on which many later icebreakers were based. In 1934 she returned to the Arctic from the Gulf of Finland where she had been employed since 1905, and she was still in regular service in 1949 when her fiftieth birthday was celebrated by the award of the Order of Lenin to the ship and to many of her crew (250).

Makarov's idea was in a certain measure vindicated shortly after his death, for the Russian Government, influenced by strategic considerations, brought into prominence by the Russo-Japanese war, ordered the construction of two icebreakers of 1320 tons displacement and 1200 h.p. which were to carry a hydrographic expedition along the Northern Sea Route. These two icebreakers were the *Taymyr* and the *Vaygach* of which we have already spoken. They were much smaller and less powerful than the *Yermak*, but they were bigger than any of the earlier craft. They were reasonably successful in doing their job, but on several occasions ice compelled them to turn back. It was clear that something more powerful was needed for work in the Arctic. The *Vaygach* sank off the mouth of the Yenisey in 1918, but the *Taymyr* was in service certainly until the second world war. Shortly after the first world war two bigger icebreakers were ordered by Russia and used in the Baltic: these were the *Petr Velikiy* of 3200 h.p. and the *Tsar Mikhail Fedorovich* of 4500 h.p. Neither of these was intended for the Arctic, and neither in fact ever went there. The first was sunk by a mine in 1915; the second changed name and ownership several times, in step with political changes in the Baltic states, but after the second world war was working for the Soviet Union, still in the Baltic, under the name of *Volhynets* (356).

The first icebreakers to be used in north Russian waters for aid to shipping, as opposed to expedition work, were those acquired by the Russian Government during the first world war for the purpose of keeping open the port of Arkhangel'sk for the import of war supplies. These vessels were almost all British-built and were brought to Arkhangel'sk between 1915 and 1918. The most powerful were the *Svyatogor* of 10,000 h.p., the *Aleksandr Nevskiy* of 7500 h.p. and the *Knyaz' Pozharskiy* of 6000 h.p. In addition there were at least fifteen smaller icebreakers built in England between 1909 and 1917; three of these were transferred from Canada. Of the considerable fleet thus assembled in the White Sea, some were sunk by hostile action, and one or two were taken away by the Allies when they withdrew in 1919. But most of the ships were still there by the time the Soviet Government gained control of the area. They were later dispersed to other ports in the Baltic, the Black Sea and the Far East where their services were needed.

A few remained in the White Sea, where they were used for seal-hunting besides port work. In some years, as we have seen, one was spared to escort the ships of the Kara expeditions, and occasionally scientific expeditions were able to charter one. Almost all Soviet icebreakers were then owned and managed by the Commissariat for Water Transport [Narodnyy Komissariat Vodnogo Transporta, abbreviated to Narkomvod], which was also responsible for all Soviet merchant shipping. Since the Arctic was visited by only a very small proportion of Narkomvod's ships, it is not surprising that little attention was paid to it. In the late 1920's, however, more icebreakers seem to have been permitted to go to the north. The search for the *Italia* survivors in 1928 was followed by expeditions to Zemlya Frantsa-Iosifa [Franz Josef Land], the Kara Sea and elsewhere; the Ob' and Yenisey traffic had icebreaker escort every year from 1929, and in 1933 icebreakers were used to escort the first freighter convoys to reach the Laptev Sea. Then by the decree of 20 July 1934 (226) Glavsevmorput' became the central authority for icebreaking services in the U.S.S.R., and obtained from Narkomvod ten icebreakers for use in the north. These ten included the most powerful: the *Yermak*, the *Krasin* (originally the *Svyatogor*), the *Lenin* (originally the *Aleksandr Nevskiy*), and the *Stepan Makarov* (originally the *Knayz' Pozharskiy*). The others were smaller: the *Fedor Litke* (originally the *Earl Grey*), the *Malygin* (originally the *Bruce*), the *Vladimir Rusanov* (originally the *Bonaventure*), the *Dobryn'ya Nikitich*, the *Truvor* (originally the *Sleipnir*), and the *Davydov* (originally the *Nadezhnyy*). Three small British-built ships joined the Glavsevmorput' fleet shortly afterwards: the *Aleksandr Sibiryakov* (originally the *Bellaventure*), the *Georgiy Sedov* (originally the *Beothic*) and the *Sadko* (originally the *Lintrose*). The last-named had sunk in the White Sea in 1916 and was raised in 1933 and put into serviceable condition. Not all these ships appear to have subsequently worked on the Northern Sea Route, judging by reports on the navigation seasons. Four ships are not mentioned in these reports: the *Dobryn'ya Nikitich*, *Truvor* and *Davydov*, which probably worked in terminal ports—possibly Vladivostok; and the *Stepan Makarov*, which seems to have worked in the Black Sea.[1] Finally, Glavsevmorput' received four icebreaking tugs from Narkomvod in 1937: *Yakutiya*, *Sigismund Levanevskiy*, *Mikhail Vodop'yanov* and *Vyacheslav Molotov*.

Soon after its creation Glavsevmorput' began to give serious consideration to the construction of new icebreakers to be built to its own design. Its newest ships were by then over fifteen years old. Further, it was essential to have a larger number of the more powerful type of vessel if the freight traffic over the Northern Sea Route were to be expanded. Already the demand for icebreaker escort exceeded the supply. Accordingly, in 1935 and 1936 four icebreakers were laid down; two at the Ordzhonikidze shipyard at Leningrad, two at the Marti yard at Nikolayev on the Black Sea (171). They were about the

[1] Fuller information on Soviet icebreakers and other ships used in the Arctic is given in Appendix VII.

same size and power as the *Krasin*—the most recent of the larger ships and itself modelled on the *Yermak*—but incorporated many improvements in design. They were of 11,000 tons maximum displacement, with triple screws driven by coal-fired triple expansion steam reciprocating engines developing 10,000 h.p. Each vessel was designed to carry three aircraft—one Dornier-Wahl seaplane and two light biplanes—and had catapult launching gear. The ships differed from earlier European and American designs in that they were intended primarily for Arctic work and therefore had greater strength built into the hull. The first of these four was to be ready for use at the end of 1936. There were delays however and in the end the first one, the *Iosif Stalin*, entered service in 1938. The *Lazar' Kaganovich* followed in 1939. Of the remaining two, one, the *Vyacheslav Molotov*, was completed at Leningrad about the time the war with Germany started, and could not therefore reach the Arctic until after the war; the other, the *Anastas Mikoyan* (which was called the *Otto Shmidt* while it was being built) left Nikolayev at the end of 1941 and reached Arkhangel'sk in 1942 by way of the Indian Ocean, the Pacific and the Northern Sea Route. It is clear from Soviet reports before the war that the performance of the *Stalin* and *Kaganovich* was satisfactory. The two later ships were of the same design but probably incorporated some minor modifications. At the time of their building these ships were the best-equipped ice-breakers afloat. *Jane's Fighting Ships* (175) mentions a further class of three ships each of 12,000 tons displacement and with 12,000 h.p. Diesel-electric engines: the *Kazak Khabarov*, *Sergey Kirov* and *Valerian Kuybyshev*. These are said to have been built in 1938–40 and to be working in the Far East. But no confirmation about these ships has been traced in any Soviet source.

Plans were drawn up for still more powerful ships. Four designs were adopted. There were three types with Diesel-electric engines, of the following sizes and power: 12,000 h.p. and 7700 tons displacement, 18,000 h.p. and 10,650 tons displacement, and 24,000 h.p. and 15,700 tons displacement. The fact that these ships displace less water in proportion to their power than their predecessors is partly a result of the change from coal-fired steam engines to the much lighter Diesel-electric engines. A 24,000 h.p. ship with turbo-electric engines was also planned, with a displacement tonnage of 16,750. These ships had evidently reached a fairly advanced stage of planning, for quite a lot of information on their design is available (364). On the basis of displacement and power of main engines, it is computed (362) that the icebreaking capacity of the planned ships is as follows: if the *Yermak* is rated at 1, and the *Krasin* and *Stalin* 1·1, then the 12,000 h.p. Diesel-electric model will be 1·1, the 18,000 h.p. Diesel-electric model 1·3 and the 24,000 h.p. turbo-electric model 1·6. The idea of the super-icebreaker even prompted a suggestion (109) that such a vessel should have a displacement of 24,000 tons and power of 52,000 h.p.; but this is scarcely likely to become a reality in the near future. There is no indication as to whether any of these have got beyond the drawing-board. Small icebreakers of the *Sedov* type, however, have also been planned, and work is said (357) to have actually been started on some of these.

Important additions to the fleet were made during the second world war by the transfer of three United States *Northwind*-class icebreakers to the Soviet Union. These vessels were built during the war and are about as powerful as the *Stalin* class but with Diesel-electric engines. The Russians gave them the names *Severnyy Veter*, *Severnyy Polyus* and *Admiral Makarov*. These vessels were lend-lease goods and therefore returnable after the war. No action was taken by the Russians until December 1949, when one ship was returned. The other two were expected to follow shortly afterwards[1]. In 1942 the Canadian icebreaker *Montcalm* was also transferred to the Soviet Union. This was an old and much less powerful vessel, built in 1904 and developing 3225 h.p.

The vessels so far mentioned do not represent the total icebreaker strength of the Soviet Union, but only, with a few exceptions, that part of it which is known to have been used in some capacity on the Northern Sea Route or by Glavsevmorput'. As we have already noted, many Soviet ports far from the Arctic require icebreakers in winter. At such ports therefore one may expect to find icebreakers whose normal work is unrelated to the Northern Sea Route. This category probably includes four ships, all over 4000 h.p., listed in *Jane's Fighting Ships* (176): the *Volhynets* already mentioned, the *Valdemars*, the *Jaakarhu* and the *Voima*. These were acquired by the Soviet Union after the occupation of the Baltic States in 1939 and the Russo-Finnish war of 1939–40. The *Valdemars* is said by Jane to be used in the Far East and the other three in the Baltic. It is of course possible that some or all of these may have been used in the Arctic during or after the war, but no mention of this has been traced in Soviet publications. It may be noted in passing that the icebreakers regularly used in the north are employed at more southerly ports during the winter.

To summarise the situation with regard to icebreakers: it is clear that the icebreakers—almost all British-built—acquired by the Russian Government during the first world war were of decisive importance in the early development of the Northern Sea Route under Glavsevmorput'. The shipping seasons of 1933–38 would have been impossible without them. The first Soviet-built icebreaker came into use in 1938, and was followed by further powerful additions to the fleet from both Soviet and foreign sources. That fleet certainly remains, after the second world war, the largest and most powerful of any country.

As far as freighters strengthened against ice are concerned, the situation is quite different. Imperial Russia possessed no fleet of such freighters. The need for such ships was limited, but even the *Kolyma*, one of the very few vessels which was regularly used in the Russian eastern Arctic, was quite inadequately protected. Between the two world wars virtually all Soviet freighters used in the north were ordinary thin-skinned vessels which spent the rest of the year in temperate or tropical seas. None of these ships was built for ice work, but many later had their ribs and hull strengthened in varying degrees. They were principally ships of 3000 to 4000 tons displacement, built to carry timber. There were two types: those built about 1927–29 and capable of 7 knots, and those built between 1935 and 1938, capable of 10 knots (337). If any of them got into difficulty with the ice, damage to the hull frequently resulted and lengthy

[1] Both were returned in December 1951 (*The Times*, 20 December 1951).

and expensive repairs had to be undertaken. Almost all these ships were coal burners. It came to be realised that conversion to oil would be a great advantage in that more space would then be available for cargo; but there is no indication that this was done, certainly up to the war.

With a few exceptions, Glavsevmorput' did not own the freighters which it employed, but chartered them from Narkomvod. It was the responsibility of Narkomvod to supply the most suitable ships available, but there were complaints that the ships provided were not at all suitable. It will be recalled that inquiries after the 1937 shipping season led to some revelations about lack of ice-worthiness in Narkomvod ships. It was impossible to do very much, however. A large slice of the budget of Glavsevmorput' was already going to the more important task of building icebreakers. It was simply not possible to undertake wholesale conversion of the Narkomvod ships. It became clear that Glavsevmorput' would have to acquire its own ice-strengthened freighters. The advantage of having a freighter which could manage as far as possible without icebreaker assistance was obvious. A sort of auxiliary icebreaker with enlarged cargo space was what was required. The first acquisition of a ship of this type was the *Chelyuskin*, which was completed in 1933 in Copenhagen. The loss early the next year of this prototype, for so it really was, was therefore a severe blow. Glavsevmorput' continued to consider the problem and evolved a design based on the *Chelyuskin*; among the modifications introduced was a hull of twice the thickness. This design was called *Sevmorput' I*. Up to 1940 only two ships of the class had been built: the *Dezhnev*, finished in 1938, and the *Levanevskiy* which was finished in 1940 but did not reach the Arctic until 1946 (193). They were built at Leningrad and had a speed of 11½ knots and a load capacity[1] of 2140 tons (361). Papanin urged the construction of more ships of this type in 1939 (235), but it is not known with what results. A design was produced for a *Sevmorput' II* ship, which was to carry 3250 tons of cargo at 12½ knots. A ship of the same size and speed but with Diesel-electric engines was also designed; this was to have a load capacity of 3880 tons (361). Vinogradov (361) mentions that some *Sevmorput' II* ships had been built, but gives no details. Two small tankers with strengthened hulls were brought into use in 1938; these, called the *Nenets* and the *Yukagir*, were built in Japan. A larger tanker for Arctic work was projected, but this was apparently still only a plan in 1945 (361). The reason why all these planned ships were so small was clearly the limitation imposed by the shallow water in many of the anchorages and in parts of the shipping routes.

During the war a number of freighters were acquired through lend-lease. Among those which sailed on the Northern Sea Route were fourteen Liberty ships, which could take a load of some 9000 long tons and were by far the largest cargo ships to have sailed in these waters. They had been strengthened in the bows, but that was all. It is therefore of considerable interest to note that all these ships apparently had successful voyages. They were scheduled to call at several quite small ports, like Ambarchik and Pevek; if, as is believed

---

[1] The Russian phrase is *chistaya gruzopod"emnost'*, which is probably equivalent to deadweight tonnage.

to be the case, they were able to discharge there and were otherwise also satisfactory, then the interesting question arises whether the new Soviet ice freighters have not been planned to be smaller than is really necessary.

As usual, there is no information on developments since 1945. The post-war five-year plan envisages expansion of the shipbuilding industry in the whole country to twice the 1940 capacity (144). Some of the projected ships must surely have been completed; for it seems reasonable to suppose that the emphasis in the Glavsevmorput' construction programme may have shifted from ice-breakers to special freighters, since only these could provide the extra speed which was essential, as Papanin made clear in 1940, if turnover was to be increased.

Besides the icebreakers and the freighters, there is another group of vessels which should be mentioned: the small craft used on the rivers and in the harbours. The powered river craft were principally strongly-built tugs, most of which were obtained from abroad. The barges and lighters were in the main of wooden construction and built locally. There were, within easy reach of the Northern Sea Route, several small shipyards which turned out barges of up to 1000 tons capacity, launches and other small craft. There was a yard at Arkhangel'sk (268), on the Ob' at Tobol'sk (198), on the upper Yenisey at Pridivnensk (Predivinsk on some maps) (163) on the upper Lena at Kachuga (297) and Peleduy (181). Kachuga alone produced metal ships; all the others built in wood. The ports at Dikson, Igarka and Tiksi were supplied with small boats by these yards. In addition, several of the 150-ton sealers used for hydro-graphic work came from the Arkhangel'sk yard.

As might be expected, the experience gained by the Russians in the manage-ment of ships in Arctic ice has led to some crystallisation of views about types of ship and features of design most suited to specific tasks. Icebreakers working in the conditions of the Northern Sea Route must be prepared to do much more than break ice. When they reach a ship in difficulty they may have to bunker it, or carry out repairs, or supply medical equipment and attention. Again, icebreakers may be the bases of scientific expeditions or the floating centres of the ice and weather service, and may therefore have to carry a considerable scientific staff; or they may have to relieve outlying polar stations and therefore have to carry freight. Subdivision of labour is necessary. Soviet writers (17,109,358) distinguish various categories of ships for ice work. Capital icebreakers [lineynyye ledokoly] are those that can lead a convoy of freighters unaided and are large enough to carry workshops and surplus fuel. The *Yermak*, the *Lenin*, the *Krasin*, the four *Stalin*-class and the three *North-wind*-class vessels are in this category. Auxiliary icebreakers are less power-ful than these but should be ideally more manœuvrable; they are often used in convoy work not as leaders but to give help to vessels near the tail of the convoy. The *Dobryn'ya Nikitich* is an example of this type. The *Litke* is some-times so considered; she is really on the borderline between this group and the preceding one. The next class, in descending order of power, is the expedition ship. These are sometimes called hydrographic ships, but should not be confused with much smaller types of hydrographic ship which are really

150-ton sealers. Expedition ships have to be capable of moving unaided, but without any ships dependent on them, in quite heavy ice. According to one authority (359), all the smaller ships sent to the White Sea in 1915–18—the *Sedov* and those like it—fall into this class. Then the last group of icebreakers are the port icebreakers, which are generally of less than 2000 h.p. Specially designed freighters like the *Chelyuskin*, the *Dezhnev* and the *Levanevskiy* are classed as icebreaking freighters. Two authorities (17,109) include the *Sedov* type with these since they have a certain amount of cargo space. The smallest and least powerful of the ships designed for the Arctic are the hunting ships which are strongly-built vessels of 150–200 tons, used for sealing and, as we have just mentioned, for hydrographic work. In the Soviet Union the word icebreaker is used to cover all these categories, even including the hunting ships on occasion. In this study the word is used to cover capital, auxiliary and port icebreakers, and expedition ships.

A good deal of attention has been paid in the Soviet Union to features of icebreaker design. The type of engine is one point at issue. All Soviet icebreakers up to the war had coal-fired steam reciprocating engines. Their disadvantages were recognised: frequent refuelling, with consequent limitation of range; and steam engines do not adequately withstand sudden temporary stoppages of the screw caused by ice, nor can they develop full power when the screw is turning at reduced revolutions through impedance by slush. In the opinion of a Soviet engineer (336) Diesel-electric or turbo-electric engines would give the best results. In addition Diesel-electric engines will carry a ship three times as far as will coal-fired steam engines on a given weight of fuel. It is clear from the icebreaker plans drawn up before the war that Soviet opinion even then favoured Diesel-electric power. No doubt the performance of the Swedish Diesel-electric icebreaker *Ymer*, built in 1932, and of the Finnish *Sisu*, built in 1939, exerted an influence here. It seems strange that none of the *Stalin*-class ships have Diesel-electric power in view of its acknowledged superiority. But probably the comparative ease with which coal from Spitsbergen could be made available in the Soviet Arctic, as against the difficulty of supplying oil there, was taken into account;[1] also, the icebreakers were going to have to carry coal anyway in order to be able to bunker the predominantly coal-burning freighters. While we are on the subject of propelling machinery, a word about propellers is relevant. The Russians have here developed two features which do not seem to be in general use elsewhere: they have found steel propellers to be more suitable than the usual bronze; and they have evolved a type with removable blades, and this has the advantage that if one blade is damaged it can be replaced, at sea if necessary, without removing the whole propeller.

The method of breaking the ice has been the subject of study. Since the first decade of this century, if not earlier, there have been two principal methods: that by which the ice is cut by the sharp bows of the icebreaker which approaches it horizontally; and that by which the bows of the icebreaker mount on top of the ice which is then crushed by the weight of the fore part

[1] The question of fuel supply for the Arctic fleet is examined in the next section (pp. 79–88).

of the ship descending vertically upon it. The Soviet Union possesses only one icebreaker of more than 5000 h.p. which uses the cutting method—the *Litke*; the icebreaking freighters of the *Sevmorput' I* and *Sevmorput' II* classes also use it (363). All the other large icebreakers crush the ice, and this method is found the most effective in close pack or anything more difficult still. Of the crushing type there are two subdivisions: with or without forward screws. It will be recalled that the *Yermak* had a forward screw at first, but it was removed because it was too easily damaged. Since that time the forward screw has been found to be a great help in ice less than one year old, such as is found in winter in the Baltic and the Pacific; but in older or thicker ice the experience of the *Yermak* has only been repeated. Recent proof of this was provided when the *Northwind*-class ships, which were built with forward screws, were acquired by the Soviet Union and worked alongside the *Stalin*-class. In the Okhotsk Sea the *Northwind* ships were found to break ice quicker and to have greater manœuvrability; but when the same ships went to the Arctic their forward screws had to be removed (15). The United States authorities had to make the same modification when they took their *Northwind*-class vessels to North American Arctic waters. Up to the end of the second world war, then, the Soviet Union found that no design for an Arctic icebreaker was more effective than that which had already been in general use for many years—the crushing method without a forward screw. The projected new icebreakers were of this design.

New methods and devices however were being made the subjects of experiments. One was based on the fact that ice can be cut by a stream of water under pressure. V. Chizhikov, an engineer, suggested the idea in 1935 and it was tested in 1938–39 at Leningrad. A stream of water 17 mm. wide at a pressure of 50–60 atmospheres (750–900 lb. per sq. in.) was able to cut a 1-m. thickness of ice in 15 minutes (69). This does not sound very impressive, but Chizhikov wrote in 1946 (32) that "in the current year Glavsevmorput' plans to equip one of its icebreakers with a hydro-icebreaking installation, which will be the first to be used for regular transportation". The pressure required was now said to be 200 atmospheres (3000 lb. per sq. in.), and Chizhikov claims that with this device a freighter of 8000 to 10,000 tons could go through ice 2 m. thick at the rate of about 2½ knots. However inaccurate these details may be, it is at least clear that the idea was not discarded out of hand, since it was still alive eleven years after it was first put forward. But it seems probable that this device could at best be used as an aid to, and not as a substitute for, the orthodox method. Another novelty in icebreaker design was suggested in 1933. This was that the stem of an icebreaker using the crushing method should be so shaped that it does not fall away steeply under the ship below the waterline in the usual way, but so that it slopes outwards, away from the ship, forming the nose of the vessel under water. The intended effect is that when the vessel rams the ice the latter is broken from its under surface. The argument in favour of the method is this. Sea ice is plastic on its under side and brittle on its upper side, particularly when the sun has thawed snow on top, allowing water to trickle through cracks and freeze again underneath into plastic fresh-water ice. When such ice is crushed from above, the brittle ice tends to absorb the blow while the plastic ice just bends, but when the blow comes from below the plastic ice has to take the full shock. Also, when sea ice is subjected to pressure from above, the water underneath will offer resistance as well as the ice, but when the pressure comes from below the only resistance is the weight of the ice. It was found after experiment in 1935–36 that sea ice can be broken from below by a force 12 % less powerful than would be required to break the same ice from above. Tests with models were carried out in 1939–40 and again in 1940–41; it was found that the vessel with the new type of bow went slower in ice than the old, but left a clearer path. Academician A. N. Krylov, a shipping expert, recommended construction of a ship of the new type. But neither Nazarov, whose paper(209) contains the information given above, nor Barabanov, who mentions the idea in a paper(15) published a year later, give any indication that work has yet started on building a ship incorporating this idea.

Work on aspects of icebreaker design was facilitated by a number of scientific studies compiled from 1935 onwards on such topics as ice pressure on ships' hulls(283,332) and relationship between the loss of speed caused by ice and the types and power of ships' engines(134). Such studies were largely undertaken by the Shipping Research Bureau of the Arctic Institute.

The technique of sailing in convoy through ice has been closely studied in the Soviet Union. The number of ships in a convoy depends upon the ice conditions, the power of the ships, and the number of icebreakers available for escort. It is always an advantage, in view of the shortness of the season, to have the convoys as large as possible, so considerable thought has been given to this. M. P. Belousov, sometime captain of the *Stalin*, has summarised the principles of convoying found most effective in a paper written about 1940(17).

There is no doubt that the Soviet Union knows much more than any other country about handling ships in Arctic waters. Even the United States, which has recently been showing great interest in icebreakers and which has four *Northwind*-class ships herself besides the three loaned to the Soviet Union, has a long way to go before she will either possess a fleet of icebreakers as large as the Soviet Union's, or will have acquired comparable experience in the technique of ice navigation in the Arctic. The Soviet Union's strength in ice-breakers has been perhaps the most important single factor in the success achieved in navigation of the Northern Sea Route. But there is certainly room for improvement, particularly in the matter of creating an adequate Arctic freighter fleet. The importance of striving after this improvement is great, for the future of the Northern Sea Route may well be closely connected with future developments in ice ship design.

(iii) *Fuel*

The availability of fuel is obviously a matter of importance to shipping on the Northern Sea Route. Before 1933 most of the ships sailing from the west were foreign and carried their own coal. Such Russian ships as were used bunkered at Arkhangel'sk or Murmansk with coal mined chiefly in the

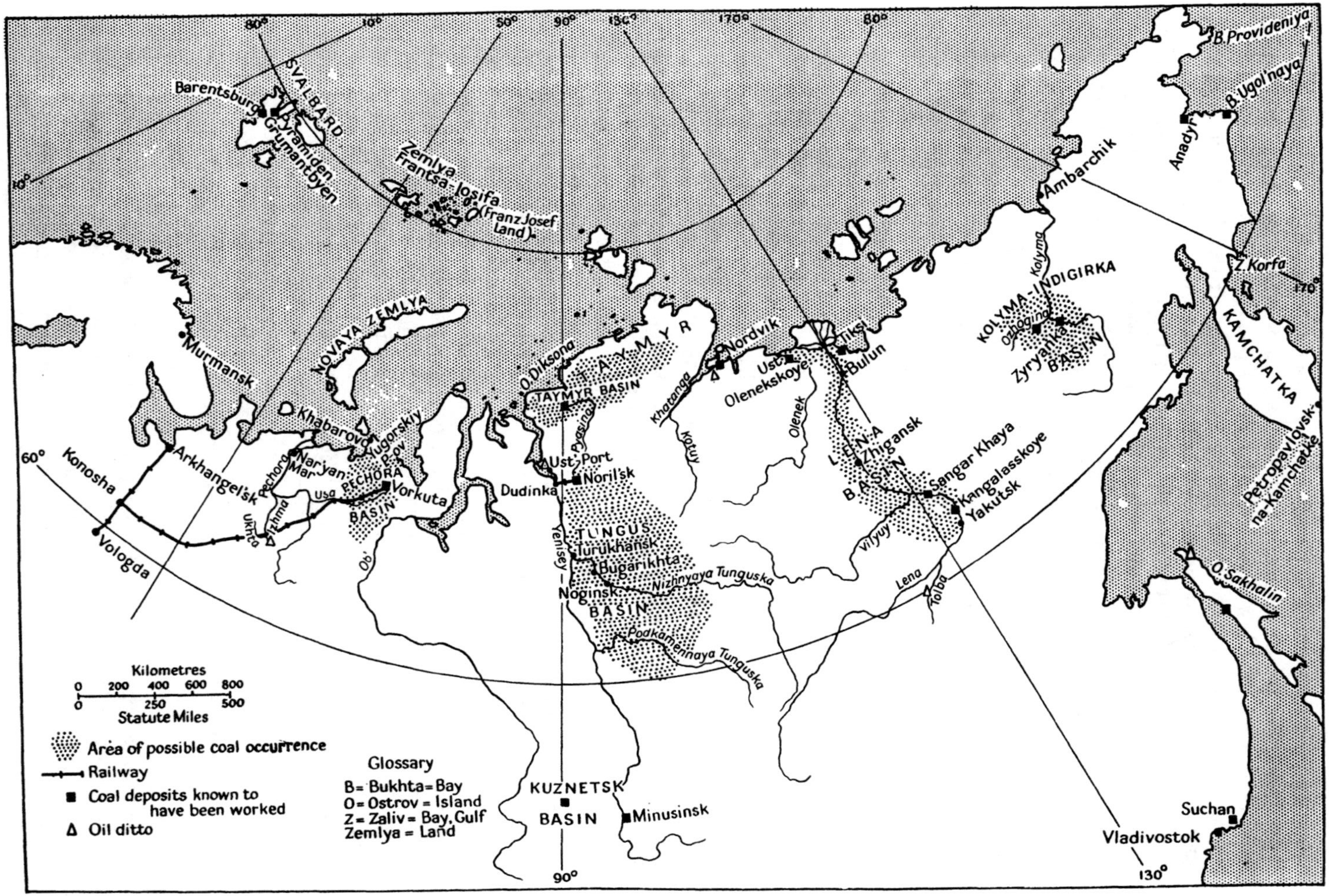

Map 6. Arctic sources of coal and oil

KAMCHATKA
Petropavlovsk na-Kamchatke
O. Sakhalin
Suchan
Vladivostok
130°
B. Providenlya
B. Ugol'naya
Anadyr'
Iskra
Ambarchik
KOLYMA-INDIGIRKA
ZYRYANSK BASIN
Sangar Khaya
Kangalasskoye
Yakutsk
Kolyma
Oyogos
LENA BASIN
Vilyuy
Lena
Toloa
Bulun
Ust'
Olenek
Olenekskoye
Norilsk
Katuy
Khatanga
TAYMYR BASIN
Ust'-Port
Norilsk
Dudinka
TUNGUS BASIN
Turukhansk
Bugarikhta
Noginsk
Nizhnaya Tunguska
Podkamennaya Tunguska
Yenisey
KUZNETSK BASIN
Minusinsk
O. Dikson
Zemlya Frantsa-Iosifa (Franz Josef Land)
Novaya Zemlya
Khabarovo
Yugorskiy
PECHORA BASIN
Mar-Yan
Vorkuta
Usa
Izhma
Pechora
Ukhta
SVALBARD
Barentsburg
Grumantbyen
Pyramiden
Murmansk
Arkhangelsk
Konosha
Vologda
60°

Kilometres
0   200 400 600 800
0   250   500
Statute Miles

Area of possible coal occurrence
Railway
Coal deposits known to have been worked
Oil ditto

Glossary
B = Bukhta = Bay
O = Ostrov = Island
Z = Zaliv = Bay, Gulf
Zemlya = Land

60°
80°
90°
60°
90°
130°
150°
170°

Donbass area in the Ukraine, or imported coal. Small quantities of coal were brought down to the mouths of the Ob' and Yenisey from the Siberian mining areas of the Kuznetsk basin and Minusinsk. Ships based on Vladivostok used coal from Suchan, near Vladivostok, or from Sakhalin. The fact that all these mines were thousands of miles from the Northern Sea Route was of course a disadvantage, but not an insuperable one. Once ships started sailing beyond the Yenisey and the Kolyma, however, it became apparent that coaling bases along the route were essential. Without them the icebreaker's range and the freighter's cargo space would both be severely limited. It also became apparent that much would be gained if local fuel were used. First, it was expensive to haul the large quantities of coal now required from the Donbass to the northern ports; even the comparatively direct route down the Ob' and Yenisey from the west Siberian coalfields is a very long distance. Second, Donbass and Siberian coal was increasingly needed elsewhere as Soviet industry expanded. The river fleets on the lower Ob' and Yenisey were similarly in need of fuel mined in the Arctic. Earlier craft burned wood, which was quite easily obtained. But the powerful tugs acquired in the late 1920's required coal, supplies of which came from the upper reaches of each river, so that fuel for the return voyage upstream had to be carried in what would otherwise have been cargo space. On the other rivers the situation was even worse since there was no direct communication with the western Siberian coalfields. Besides the sea and river fleets, there were other potential users of coal mined in the Arctic: in particular the industrial undertakings for which Glavsevmorput' became responsible. In face of this large demand, supply was limited to coal from one or two Arctic deposits which had already been worked on a small scale. It was known however that several more existed; steps were accordingly taken to study more closely the potentialities of these, and geological expeditions were sent out to look for new coal-bearing areas, and also to examine the possibility of finding oil.

In the western sector it was necessary to supply the two terminal ports of Murmansk and Arkhangel'sk, and a coaling station in the Kara Sea. The obvious site for this last was Ostrov Diksona, and it was selected. The most important local coal source for these three in the early years of the existence of Glavsevmorput' was Svalbard, the group of islands north of Norway. V. A. Rusanov, the Russian explorer, staked claims for coal in Vestspitsbergen, the main island of the group, during his expedition of 1912. His claim at what is now called Grumantbyen was taken over by Russian émigrés, who formed the Anglo-Russian Grumant Company and started working the claim in 1919. By 1924 annual production had reached 9000 tons, of sufficiently good-quality coal for use in ships. Then the Soviet timber trust, Severoles, which required coal for the ports of Arkhangel'sk and Murmansk, bought shares in the company. Production rose to over 20,000 tons a year. In 1931 the remaining shares were bought up by the Soviet Union, and in 1932 the Dutch mine at Barentsburg was purchased. These two mines, together with claims in the Pyramiden area to the north-west and Bohemanflya to the north, were run by a new trust called Arktikugol' (326). The coal was largely used at Murmansk by Soviet trawlers. It was clear that production would have to be increased enormously if the ships bound for the Northern Sea Route were also to be supplied. The number of miners was increased to 2000 in the winter of 1932–33, and in the next two years shock workers from the Donbass were sent there. This had the effect of raising production until it reached and probably exceeded the level of 400,000 tons a year in 1936 (79). Russian sources (325) indicate that production remained at about the same level until 1941, but Norwegian statistics (216) show a decline after 1937 to about 270,000 tons in 1940. The Barentsburg mine was used as a training centre for Arctic miners. Preparatory work was done on the Pyramiden site, but no coal was got out before the war. The result of the great increase in production in the early 1930's, however, was that by 1939 the western coaling bases of Glavsevmorput' were very largely supplied by Spits- bergen coal.

But there were disadvantages in this coal. The seams at Barentsburg, the largest mine, were thought to be nearly worked out (185). From the point of view of economics, Spitsbergen was after all quite a long way even from Mur- mansk and the necessary transport had to be provided. But more important, the position of Spitsbergen made it vulnerable in case of war. By 1939 there- fore it was clearly imperative to develop other coalfields on the Soviet main- land to replace Spitsbergen output. With this end in view, speedier development of the coalfield on and near the Vorkuta, a tributary of the Pechora, was undertaken. It had been known for some time that coal occurred here, and small-scale mining started in 1931 (189). But the remoteness of the region made it almost impossible to get the coal away until some forty miles of railway were built from the mine to the nearest navigable river, the Usa. This was done by 1939, or possibly earlier, and the coal was taken down the Usa and the Pechora to Nar'yan-Mar, whence it could be shipped to Arkhangel'sk or Murmansk (184). The reserves of this coalfield were stated (249) by the director of the mines in 1946 to be 36,000 million tons—rather under half the size of the Donbass. Although this was a good deal smaller than earlier estimates, the size of the deposit and the quality of the coal clearly had potential national importance. In order to avoid the transport limitations imposed by the freezing of the Pechora and the sea, work was started on a railway which was to run from Vorkuta to join the railway network of the Soviet Union at Konosha, between Vologda and Arkhangel'sk—a distance of over 1000 miles. The route passed through another newly discovered coal-bearing area on the Ukhta river, 400 miles south-west of the Vorkuta mines. Mining is said to have started here in 1941, but no output figures are available. The line was finished in 1943, and since then has carried the greater part of the coal. Output in the Vorkuta area grew extraordinarily quickly, thanks partly at least to drafting a large number of prisoners to the mines. Production in 1945 was said to be eleven times that of 1940, and 1945 production was to be tripled by 1950 (320). The post-war five-year plan mentions (143) that new pits with a capacity of 7,700,000 tons are to be brought into production by 1950. The increase in production during the period of the plan is therefore not likely to be less than this figure; so production in 1945 was probably at least 3,800,000 tons, and that envisaged for 1950 11,400,000. Of course it may have been considerably greater. An English source (266) gives output for 1941 as 1,000,000 tons; this would put 1945 output in the region of 10,000,000, and the plan for 1950 at about 30,000,000 tons. But not too much reliance should be placed in this. Coals from the Pechora basin have been chiefly used to supply Leningrad and other industrial centres in the northern part of European Russia and of the Ural mountains. But presumably they were also used for bunkers at Arkhangel'sk and Murmansk. It was intended (395) in the early 1930's to build a railway from Vorkuta northwards to reach the sea probably at Khabarovo. This would have made the coal even more readily available to ships using the Northern Sea Route, providing a new coaling station between the White Sea and Ostrov Diksona. But although an English source (219) states that this line was completed in 1943, Soviet maps published since that date do not mark it.

The supply of Spitsbergen coal, which came to an abrupt end in 1941 when the war situation made it necessary for the mines to be rendered useless and the miners evacuated, was thus more than replaced. Nevertheless after the war Soviet miners returned to Spitsbergen. By 1948 there were about 2500 Russians wintering in the three settlements (1). Barentsburg and Grumantbyen were being repaired, and Pyramiden was starting to produce coal. Soviet production however had only reached an estimated 80,000 tons in that year (217). A speedy return to the peak pre-war level of 400,000 tons a year seems unlikely.

Supplies of coal for the western terminal ports should now be adequate, even for the increased number of ships which require it. No doubt the coaling station at Ostrov Diksona could also be supplied, as it had been earlier. But plans had been advanced before the war for supplying Ostrov Diksona from closer sources.

One of these sources was Noril'sk, about fifty miles east of the lower Yenisey settlement of Dudinka. Coal had been noticed here by Middendorf, the scientist and traveller in Siberia, in 1843. It was first used on a river boat when a Dudinka merchant mined 35 tons of it in 1894 (218). In 1905 500 tons were mined for the use of river craft meeting the large Kara expedition of that year (201). A. I. Vil'kitskiy said that the coal was as high in quality as English coal (218). The area was surveyed in 1919–20 and Komseverput' sent a reconnaissance party in 1921 to study the possibility of starting mining on an industrial scale. During the course of the next few years a number of other minerals were found in the area, notably nickel, zinc and copper, and the Northern Polymetal Combine was formed to develop all this. Mining was not started until about 1930, and when it did start the metals were given priority over the coal. Access to the area was not easy. We have already seen how convoys of river ships were sent down the Yenisey and up the Pyasina with supplies for Noril'sk. It was not until the Noril'sk-Dudinka railway, which had been under discussion since 1920, was completed in 1937 or 1938 that the coal acquired anything more than local importance. In 1938 Noril'sk coal reached Ostrov Diksona for the first time (178) and in 1939 ships leaving Igarka for the Kara Sea bunkered at Dudinka. Probably Noril'sk coal has been used regularly at Ostrov Diksona ever since. But it is not possible to say whether it satisfies the whole demand, or whether coal has also to be brought from elsewhere, since no production figures are available.

It was thought by some however that Noril'sk coal was not the answer to all requirements in the Kara Sea area. A series of geological expeditions working in Taymyr between 1935 and 1939 established the presence of a coalfield between the Pyasina and the Yenisey and the hypothesis was put forward that this was one end of a coal-bearing belt stretching north-eastwards across Taymyr (333). The Yenisey-Pyasina coalfield is very much closer to Ostrov Diksona than is Noril'sk. Certain writers (187) are much in favour of using this coal, which they claim is of better quality than that from Noril'sk, as well as being closer. But others (195,335) state that although there is good coal in the new field it is largely inaccessible, while the accessible parts of the field near the coast contain only what is described as "highly metamorphosed" coal. In 1940 100 tons of coal from one of the coastal seams (near the river Krest'yanka, which reaches the sea about 45 miles south of Ostrov Diksona) was used experimentally by a ship, and was found to be not so good as Noril'sk coal (334). Towards the end of the war however mining started on the left bank of the Pyasina about fifty miles from its mouth. The coal was said to be good enough for ships, but there is no information on the extent to which it was mined or used (62). Whether Pyasina coal will replace Noril'sk coal as the source of supply for Dikson remains to be seen. In any case the Taymyr coalfield is a useful reserve.

There are other coalfields which, from their location, might provide fuel for shipping in the Kara Sea area. The Tungus basin is the name given to a very large area between the Yenisey in the west and the upper reaches of the Vilyuy in the east, stretching from lat. 57° N. to 71° N. Coal occurs in a number of places in this basin. The Noril'sk coalfield is at the northern end, and there are others on the Podkamennaya Tunguska and the Nizhnyaya Tunguska, right tributaries of the Yenisey. Some deposits, particularly those at Bugarikhta and Noginsk on the Nizhnyaya Tunguska, have been worked sporadically, and coal from them was used by the Yenisey river fleet in the early 1930's (38). But about 1934 production stopped and apparently had not by 1939 started again. Possibly the reason was the poor quality of the coal, which had a low volatile component. It is also true that a feature of the whole region is the breaking-up of the coal measures by intrusive rocks of younger age, and this complicates mining (182). Another coalfield, much closer to the shores of the Kara Sea, is situated on the east side of Yugorskiy Poluostrov. It was discovered in 1932 and studied in greater detail in 1933 and 1936–38. The coal has a high ash content and low calorific value, which rule out its usefulness for ships (47). In any case the area is not easily accessible from the sea, for although Baydaratskaya Guba is close at hand its waters are shallow and stormy. There is little likelihood of this coalfield being developed in the foreseeable future. Low-quality coal has also been found on Zemlya Frantsa-Iosifa [Franz Josef Land], but it is even less likely that it will ever be thought worth while to overcome the difficulties imposed by the elements in order to mine it.

The Laptev Sea needs a local source of fuel more than any other sea on the

Northern Sea Route by reason of its central position. The first usable deposit to be found within a reasonable distance of the Laptev Sea was that at Sangar-Khaya, about 750 miles up the Lena. This was first surveyed in 1913 and first worked in 1928. The coal, which is between brown and stone coal, was used for the river fleet and after 1933 was sent down to Tiksi. By 1936 output reached 23,000 tons a year (191). (Another source (374) puts production at 33,500 tons in 1927.) The coal was not wholly satisfactory for ships; it may be recalled that Sangar-Khaya coal was blamed for causing some ships to winter at sea in 1937. Nevertheless Sangar-Khaya continued to supply Tiksi at least until the war.

Coal is known to occur at other points on the Lena. At Bulun, which is only 200 sea miles from Tiksi, the ash content is too high. Brown coal has been found at Zhigansk, at a number of points in the basin of the Vilyuy, a left tributary of the Lena, and at Kangalasskoye, on the Lena 25 miles below Yakutsk. Only the last-named deposit is worked, and the coal is used for the local needs of the town of Yakutsk.

In 1941 a 15-m. seam of brown coal was discovered on the river Sogo, only eight miles from Tiksi. Small quantities had been found here earlier but had not been thought worth further investigation. Although the coal is brown, its very convenient situation and the size of the seam have led to discussion on the possibility of making the coal into briquettes for use in ships (41).

Coal has been found at four other places in the Laptev Sea region. The most promising seems to be that on the lower Olenek. Within sixty miles of Ust'-Olenekskoye two deposits of boghead coal were found by expeditions working in 1933 and 1938–39 (352). The coal was thought to be good enough for refinery into liquid fuel, and in 1944 an experimental distillery was set up at Tiksi (99). The post-war five-year plan calls for further work on this (75). At Nordvik brown coal was used for local needs in 1934–36 and again after 1942. It seems unlikely to be used for anything else. When the *Krasin* was forced to winter here in 1937–38 this coal allowed the ship to reach Tiksi, but the quality was very poor. On the Kotuy, a right tributary of the Khatanga, there is known to be coal of a sort; it was investigated in 1939 with a view to its use by ships calling at the Khatanga, but the quality is not high.

East of the Laptev Sea the principal source of coal is the neighbourhood of the Zyryanka, a left tributary of the Kolyma. The deposit was discovered in 1891, and mining prospects were studied in 1933–34. The Kolyma river fleet, belonging to Dal'stroy, was the first customer. From 1937 or 1938 coal was available at Ambarchik, at the mouth of the Kolyma, for bunkers, although the quality does not appear to have been very high (296). In 1939 about 6000 tons of Zyryanka coal was shipped to Pevek and Bukhta Provideniya (74). There is another coalfield about fifty miles from the Zyryanka, on the Ozhogina, and this is probably also being exploited (296). It is likely that the Dal'stroy mines some 250 miles to the south are now supplied, partly at least, by these two coalfields.

No other coal within reach of the Northern Sea Route between Tiksi and the Pacific had been found by 1939. The anchorage at Bukhta Provideniya on

Bering Strait had been used as a coaling station continuously since 1933 and sporadically before that. The coal was brought mainly from Vladivostok. Glavsevmorput' studied the possibility of supplying Bukhta Provideniya from closer sources. Coal occurred, and was used for local needs after 1929, at Anadyr'. An expedition worked in the area in 1933–34, but found that of the two coalfields here and several others between Anadyr' and Bukhta Provideniya, none contained coal of high enough quality for ships (186). In 1935 however a much more likely area was surveyed at Bukhta Ugol'naya, on the coast 150 miles south of Anadyr'. The coal here was suitable for ships, and small quantities were mined (187). Early in 1940 the formation of an organisation called Bukhtugol'stroy to develop a mining centre here was announced (229).

Farther south, coal was found at Zaliv Korfa in north-eastern Kamchatka. Coal was needed in this locality to fill the large gap between Bukhta Ugol'naya and Sakhalin; but it is said (37) that this coal was too poor in quality for ships. As it is, the intermediate base at Petropavlovsk-na-Kamchatke is normally supplied from Sakhalin or Vladivostok.

To summarise the foregoing: in 1932 the only coal mined in the Arctic that was available for ships was Spitsbergen coal. By 1940 this was augmented by the discovery or enlargement of mining centres at Vorkuta, serving the western terminal ports; at Noril'sk, serving Ostrov Diksona; at Sangar-Khaya, serving Tiksi; at Zyryanka, serving Ambarchik; at Bukhta Ugol'naya, serving Bukhta Provideniya. Vorkuta and Spitsbergen were incomparably the largest of these, between them accounting for at least 90 % of the total Arctic output. In 1939 the Head of Glavsevmorput' announced (237) that 78 % of the coal used by ships sailing on the Northern Sea Route came from Soviet Arctic sources. It is probable that most of the remaining 22 % is accounted for by coal taken aboard at Vladivostok.

Of developments since 1940 it is difficult to say anything. If the Vorkuta coalfield could reach an output of several million tons a year in such a short time, could the same sort of thing have happened elsewhere? In fact lend-lease goods and equipment were sent during the war to the mines at Sangar-Khaya, Zyryanka, Bukhta Ugol'naya and Tiksi, and to the boghead refinery at Tiksi. This seems to indicate expansion of the first three, and confirms that serious attention was being paid to the deposits on the Sogo and Olenek. But on the whole it is unlikely that expansion comparable to that at Vorkuta took place anywhere else in the north. The Vorkuta mines were enlarged to meet a national need which far exceeded in importance the needs of the Northern Sea Route; for Vorkuta coal was expected to replace not merely Spitsbergen coal in the White Sea and Barents Sea ports but Donbass coal in the northern industrial areas. There was not the same urgency about the development of the other regions. Also, many of the other coalfields contain coal of a quality which is either too low for maritime or industrial use at all— as Bulun or Anadyr'—or else has only been used because nothing better was available within a wide radius—as Sangar-Khaya. Indeed, P. P. Shirshov, a Deputy Head of Glavsevmorput', complained (285) in 1939 that the department had no good fuel sources; but it must be remembered that at this time

Glavsevmorput' controlled mining undertakings in a limited area only. The requirements of Glavsevmorput' in coal quality are given (183) as follows: 15–35 % volatiles; calorific value not less than 7000; not more that 12 % ash, 5 % moisture and 4–5 %[1] sulphur; brown coal to be used only as an exception. Of all the coals mentioned in the preceding paragraphs, the following were known in 1939 to meet these specifications: those from Spitsbergen, Vorkuta, Noril'sk, Zyryanka, Bukhta Ugol'naya, some seams in the Nizhnyaya Tunguska area and in south-western Taymyr, and possibly the Tiksi brown coals when made into briquettes.[2] Production has in fact started, though sometimes on a very small scale, at all these places except the Nizhnyaya Tunguska, where there are difficulties, as we have seen. At the lowest estimate, it may reasonably be assumed that supply was in 1949 near to meeting demand.

So much for coal. The alternative fuel was oil. In the 1930's coal was the more essential because the large majority of ships sailing in the north were coal-fired. Any oil that was required reached the western terminal ports from the principal Soviet oilfield at Baku on the Caspian Sea, and reached Vladivostok from Sakhalin, where the Russians had been getting oil since 1928. The advantage of converting ships to oil began to be realised, however, and with it arose the necessity for investigating possible local sources of supply.

Geological expeditions, working after 1932, indicated the presence of possible oil-bearing areas at three localities in Novaya Zemlya; in the region of the rivers Ukhta and Izhma, tributaries of the Pechora; at the mouth of the Yenisey; in the vicinity of Turukhansk; at Nordvik; on the upper Olenek; on the river Tolba in Yakutiya; and in western Kamchatka.

The Novaya Zemlya deposits were not considered worth exploring further. The oil of the Ukhta region had long been known to exist. Sidorov tried to exploit it in the 1870's, and claimed that it was known and even used in the time of Peter the Great. The area was studied and drilling started in 1938. The results were evidently successful. It is alleged (266), although the report may well be inaccurate, that a refinery with a capacity of about 5,000,000 tons a year was completed in 1942. This oil, however, like the Vorkuta coal, is only likely to be available for Glavsevmorput' after more pressing industrial needs have been met. At Ust'-Port at the mouth of the Yenisey methane was found. Survey work was done and deep boring started in 1939 (65). By 1940 no oil had come to the surface, and a later report (64) suggests that not too much should be expected of the area for geological reasons. The Turukhansk district was thought to be promising, but by 1940 no detailed study of it had been made (256). The same was true of the Kamchatka area. Nordvik was thought to be a very promising place. In 1927 the theory was first put forward on the basis of the similarity of the Nordvik salt domes to those in the Texas oilfields. Work was started on the site in 1933. Bore holes were sunk at a number of points in the years that followed, but without success (116). Nordvikstroy, the trust responsible for developing the mineral resources of the area, was in fact

---

[1] Presumably a misprint for 0·4–0·5 %.

[2] A list of proximate analyses of all coals mentioned in this section is given in Appendix VIII.

dissolved in 1940 (see p. 60). Interest revived in 1943 when oil was struck on Poluostrov Yurung-Tumus, and eighty tons of oil were obtained from one well in nine months. The maximum flow was 950 litres in 21 hours, obtained by pumping (137). Some of the other wells which had yielded a little were closed down. But in the opinion of one writer (138) at least, the results at Yurung-Tumus, though very poor by normal standards, justify further boring; but it is not known whether this was done or not. On the upper Olenek, or more exactly its tributary the Kenelikan, the presence of oil was definitely established, but nothing further had been done by 1946 (64). On the river Tolba, or Tuolba, a right tributary of the Lena above Yakutsk, a very small amount of oil—under a litre a day—was yielded, but no more information than this is available (257).

Thus the only practical result of the work done between 1932 and about 1945 was that oil was obtained from Ukhta. It is to be presumed that this oil was made available to ships on the Northern Sea Route but only through the terminal ports. However, the drilling at Nordvik, Ust'-Port and on the Tolba provided useful experience in the technique of obtaining oil which is lying under or in permanently frozen soil. On the Tolba the oil came up, though in very small quantities, through a permanently frozen layer. At Norman Wells in Canada the same problem has been successfully solved. But at Nordvik the permanently frozen layer extends downwards for 540 m. (136), and the well which gave the largest flow struck oil at 116 m. (137). The solution, even if incomplete, of this problem is an important achievement. The production of usable quantities of oil for use by ships and industry in the Arctic was one of the tasks set by the post-war five-year plan (75).

The fact that preliminary work at Nordvik continued for such a long period led many to assume that large-scale production of oil was only just round the corner, and therefore to urge early conversion of ships to oil fuel. That this conversion was not carried out tends to show that availability of local fuel was regarded as a *sine qua non*. If this is true, it would appear that there is likely to be further postponement.

5. SCIENTIFIC SUPPORT

We have outlined the material equipment of the Northern Sea Route with ports, ships and supply of fuel. An important feature of this work remains to be described. The scientific approach to all problems is one of the principles of Marxism. Therefore an extensive scientific programme was undertaken as an essential accompaniment to the work of construction and development. We have seen already what was done in this way before 1932. After the creation of Glavsevmorput' there was a further marked increase in the amount of work undertaken, and also an enlargement of its scope. Hydrography, hydrology and meteorology remained the principal fields of study directly related to the sea route, followed by terrestrial magnetism, gravity, actinometry and study of the ionosphere and the propagation of radio waves. On the basis of hydrological and meteorological observations and of the study of sea ice, the very important science of ice forecasting was developed. Field work was done as before by scientific expeditions and polar stations, both of which were increased in number. An outline of these two activities provides a necessary background for an examination of the scientific advances that were made, and of their effectiveness.

(i) Expeditions and polar stations

The advantage of putting a team of scientists on board an icebreaker and sending them to work during the summer in otherwise inaccessible places had been realised in the 1920's. But it had been difficult at that time for the scientists themselves to arrange this. The Arctic Institute had been able to charter an icebreaker on several occasions, but this imposed a severe strain on resources. The situation changed when Glavsevmorput' became the owner of a number of these ships. One or two of the *Sedov* type were used each year for primarily scientific voyages; and teams of scientists were generally permitted to work on board the other icebreakers, since wherever the ships' assignments might take them, there would, in the early stages at least, be plenty of scope for useful work. In this way a large number of scientists were put in the field. There were also a number of small boats working whole-time for the Hydrographic Administration of Glavsevmorput'. All these expeditions, which assumed a routine character from 1933 onwards, increased knowledge of the seaways normally used by convoys. In addition there were some expeditions which deserve special mention.

There were three consecutive expeditions in the *Sadko* in 1935, 1936 and 1937. This ship was one of the small icebreakers which frequently made purely scientific voyages. This series of three were called "high-latitude expeditions", because their range extended farther north than usual. The first, in 1935, covered the north part of the Kara Sea, between Zemlya Frantsa-Iosifa [Franz Josef Land] and Severnaya Zemlya (400). The second was to go to the north part of the Laptev Sea, especially the region north of Ostrova Novosibirskiye; but assistance had to be given to ships in the west part of the Kara Sea, and as a result of the delay caused by this the expedition went in the end to Zemlya Frantsa-Iosifa again (270). The third high-latitude expedition was able to go to the area that the second had been intended to study (380). It stayed longer in the area than was planned, for the *Sadko* was ordered to help some ships in the north-west part of the Laptev Sea and was herself caught in the ice. She drifted northwards together with the *Malygin* and the *Sedov* until the following summer when she was released. These three expeditions of the *Sadko* were called "complex"—that is, specialists in many branches of science were carried. But the chief contribution was to knowledge of the behaviour of the sea and ice in the waters north of the traffic lanes; and this was important because the possibility of using the northern variant was then under discussion (see p. 38).

In so far as all the seas crossed by the Northern Sea Route are really bays of the Arctic Ocean, study of that ocean beyond the limits of the individual seas may be relevant to problems of navigation. This is especially true if an overall picture of meteorological or sea ice conditions is to be obtained. Three Soviet expeditions during this period provided information—valuable because such information was almost non-existant hitherto—on these subjects in various parts of the central polar basin.

The North Pole drifting expedition was the most spectacular of the three. It was decided to set up a drifting polar station on the ice at the North Pole. Four men—I. D. Papanin, the leader, Ye. K. Fedorov and P. P. Shirshov, scientists, and E. T. Krenkel', wireless operator—were flown to the North Pole with their stores and equipment in May 1937. Camp was set up, and the party drifted in the next nine months to a point off Scoresbysund in East Greenland, where it was rescued from its then rapidly disintegrating ice floe. During the period of drift a full programme of observations was carried out by the two scientists. Their results (344) contributed to the formulation of theories of meteorological processes and of ice drift in the central Arctic. Hitherto the only observations on which it was possible to base such theories were those made by Nansen's party in the *Fram* during her drift across the polar basin in 1893–96.

A set of observations (13, 28) providing an even more interesting comparison with Nansen's was obtained by those on board the *Sedov*, which drifted from 1937 to 1940 on a course similar to and rather more northerly than that of the *Fram*. It will be recalled that the *Sedov* was trapped in the ice of the Laptev Sea in 1937 and could not be extricated the following year. The drift that followed was therefore unintentional and for that reason the ship was not equipped as it should have been in the matter of stores, scientific instruments or personnel. There was only one scientist aboard—V. Kh. Buynitskiy, a hydrographer. The observations therefore are not so numerous as Nansen's, but nevertheless they are very valuable, especially as the course of drift crossed that of the *Fram* on several occasions and thus gives interesting opportunities for comparison (13).

The third expedition was a series of flights to the so-called "Pole of inaccessibility" in April 1941. Landings were made on the sea ice at three places in the general region of lat. 80° N., long. 180°, that is, about midway between Bering Strait and the North Pole. From three to six days were spent on the ice at each landing place, and various observations in hydrology, meteorology, terrestrial magnetism and the drift of sea ice were made by the three scientists who accompanied the aircraft (382).

These three expeditions thus provided some fundamental data from three different parts of the central polar basin, each of them relevant in certain contexts to problems of the Northern Sea Route.

While expeditions were able to fill important gaps in knowledge of the area, it was the polar stations which provided the essential background information. There were 24 coastal polar stations in existence at the end of 1932 (see p. 35). The network was rapidly expanded. The exact number of coastal stations at any given time is hard to compute since there are a number of borderline cases—stations a short way inland up rivers, subsidiary or part-time posts manned from nearby stations, and so on—which are included by some authorities and not by others.[1] According to one source (338), by 1941 there were 77 polar stations in use, and the same number was said (58) to have been maintained throughout the war. Another source (130) quotes 62 as the number of coastal stations functioning in 1945, and according to the International Meteorological Organisation (157) there were 69 coastal stations in 1948. Whatever the exact number, it is clear that the Kara Sea area was still much the best served, having roughly half the total number of stations. But the proportionate increase since 1932 in stations in the Laptev, East Siberian and Chukchi Seas was greater; where there had been six there were now (after the war) between thirty and forty.

All stations, with the exception of those in Zemlya Frantsa-Iosifa, were within a few miles of the shipping lanes. There was talk of building stations during the third five-year plan on the northern side of Severnaya Zemlya, Ostrova Novosibirskiye and Ostrov Vrangelya in order to study the possibility of routing ships north of these islands (177). But as far as is known nothing was done during the course of the plan beyond the establishment of a station in 1941 on Mys Molotova, the most northerly point of Severnaya Zemlya, and this had to be closed after three months because of the difficulty of relief. Thus there remained outside the network many areas of possible use to shipping; but nevertheless the network had been greatly strengthened and was by about 1940 considered adequate—evidently in spite of the uneven distribution between east and west—for the needs of the normally used shipping lanes. According however to a Soviet newspaper report in April 1946 (88) the network was to be further extended during the post-war five-year plan; but since specific mention was made of projected stations covering northern waters at Mys Molotova and Ostrov Bennetta, it is possible that the expansion was limited to putting into effect the unachieved earlier plan of servicing the northern variant.

The stations were not all the same size. They ranged from large establishments like Ostrov Diksona where a staff of about 120 winterers were engaged in many different types of scientific work, to part-time stations such as Ostrov Tyrtova in 1940, where two or three men were sent during the navigation season to broadcast ice and weather reports. The majority of stations were manned by four to six winterers who were normally relieved each summer. Generally about half the staff were scientists. All stations were equipped with radio (though this was not always in working order). The radio network extended upwards from the out-stations through progressively larger regional stations to so-called radio centres. At first there was one of these in Moscow and another in Yakutsk; one was completed at Ostrov Diksona in 1935, another at Mys Shmidta in 1940, and there is probably another at Tiksi.

All coastal stations performed the basic functions of keeping a meteorological and ice log and doing "coastal hydrology" (principally tidal observations). After the basic functions, study of the upper atmosphere comes next in frequency of performance: out of 47 stations in service during the winter of

---

[1] As full a list as possible is given in Appendix IX.

1938–39, twenty were using pilot balloons and seven, radiosondes. At the same period magnetic observations were being made at five stations, four in the western sector; actinometrical work was being done at four, of which three were in the western sector; and study of the ionosphere was started during that winter at one (66). All these stations were larger than average. As the total number of stations grew, so did the number of those doing specialised work. By 1944 two more stations were engaged in magnetic work, and one more in study of the ionosphere (231).

Many stations were in direct radio contact with aircraft and ships in the vicinity, and were thus of immediate practical use to navigators. They were also used as centres of scientific expeditions, often organised by the station personnel; and as first-aid posts and bases for rescue expeditions. But their principal usefulness was as collecting points for scientific data.

(ii) *Work and results*

Space does not permit a review of the work done in all branches of science studied in the Arctic. We must restrict ourselves to consideration of how much was done, and how effectively, in those subjects which contributed most to the development of navigation—hydrography and hydrology, meteorology, the study of ice and ice forecasting, and terrestrial magnetism.

(a) HYDROGRAPHY AND HYDROLOGY

Hydrography is the branch of science most closely concerned with the safety of shipping, and in 1933 the gaps in hydrographic knowledge of the Northern Sea Route were very large. Although hydrographic work was done both before the Revolution and in the 1920's (see pp. 14–16, 32–34), the Kara Sea was the only one of the four north Siberian seas for which sailing directions and adequate charts existed in 1933. These, furthermore, were almost entirely concerned with the southern and eastern parts. There was an enormous amount of work to be done if the safety of shipping was to be secured along the whole length of the seaway.

Under Glavsevmorput', hydrographic work was generally performed by parties sailing in about half a dozen small ships belonging to the Hydrographic Administration of Glavsevmorput'. During the summer these parties charted offshore waters and set up navigational signs, starting in the areas most dangerous to shipping. Occasionally ships wintered in the locality in which they were working in order to study winter conditions, about which very little was known. Such parties visited the island groups on the east and west coasts of Taymyr, Severnaya Zemlya and Proliv Lapteva between 1936 and 1946. Charts of rivers were prepared by land-based parties which were generally in the field for a year at a time.

The study of marine hydrology is complementary to hydrographic work. Normally hydrologists accompanied the "complex" expeditions, worked from icebreakers and freighters, and were assigned to polar stations. While it is difficult to assess the exact numbers of ship-borne hydrological parties, it is probably true to say that there were never less than ten working each year. They worked in all four seas, and particularly in the straits between the seas. The first wholly hydrological expedition, as distinct from the parties aboard ships doing other tasks, was carried by the *Nerpa* in 1936 and worked in the Kara Sea, studying especially the open-sea currents. After 1938 hydrologists worked from the ice patrol boats (of which more will be said later), and continued to do so during the war.

The results of the work were briefly as follows. By the end of 1937 over seventy charts had been issued (343), but there were still no sailing directions other than those for the Kara Sea, published in 1930. In 1938 sailing directions for the Laptev Sea and the Chukchi Sea were published, together with a new edition of the Kara Sea pilot; that for the East Siberian Sea followed in 1939. It is true that there had been some information on the three eastern seas before publication of the pilot books, in what were called *Materials for the Pilot* [*Materialy po lotsii*]; but the information had been scanty and not of great value. Supplements were later published to keep the four books up to date. A large number of navigational aids, from lighthouses to unlit buoys and markers, were built, principally in estuaries and the vicinity of ports. According to the decree of 25 January 1941 (309), three main hydrographic bases were established at Arkhangel'sk, Tiksi and Bukhta Provideniya; and pilot stations were set up on the seven large rivers of north-eastern Siberia—the Khatanga, Anabar, Olenek, Lena, Yana, Indigirka and Kolyma. The Ob' and Yenisey are not mentioned since they were then outside the Glavsevmorput' system, but they had had a river pilot service for many years. The results of the hydrological work were published in a series of papers on the basis of which reference volumes were compiled—tidal and current atlases, for instance, and exhaustive surveys of each sea (72). Relevant information was included in the sailing directions.

Let us now consider the effectiveness of this work. One writer (2) describes the hydrographic work done before the Revolution as a preliminary reconnaissance, and that of the second five-year plan (1932–37) as a detailed reconnaissance; the basic achievements of the latter were the establishment of the general direction of the route and of its main variant courses, and the determination of the general character of those parts of it which were most difficult to navigate. This seems a fair statement. During the second period there was certainly a real attempt at planning the work of the various parties. The results were not more striking, in spite of the expenditure of great effort, for two reasons: first, the task itself was enormous; and second, a certain amount of effort was wasted. The first point is obvious, but was apparently not always fully appreciated by executives of Glavsevmorput' who were constantly accusing the Hydrographic Administration of lagging behind. But it is true that there was some waste of effort. The collation and working-up of results was a slow and erratic process and all available data were not always made use of in compilations. This helped to account for the late appearance of the sailing directions for the eastern seas. One of these volumes, incidentally—that for the East Siberian Sea—was thought very poor by a reviewer (53) well acquainted with the sea in question. The recording of hydrological observations was not always reliable: a thermometer at Ostrov Diksona was found to have an error of 8° C. (14·4° F.) (29). There is reason to suppose that many observations were rendered useless through similar if less extreme faults. There was also a certain confusion arising from the fact that several different authorities were concerned in the direction of hydrological work at polar stations and at sea. The task during the third five-year plan was to set these faults right and build on the foundations already laid. It is clear that the need for more detailed hydrographic information was by this time urgent—as indeed might be expected—for Papanin (235) in his speech to the Eighteenth Communist Party Congress in 1939 places it first on his list of priorities. Accordingly more resources were allowed to the Hydrographic Administration, which acquired more ships and planned more expeditions. How successful these measures were cannot be accurately assessed. But there continued to be complaints that the Hydrographic Administration could not keep up with demands made upon it; and in 1941 more hard words from Papanin (233) make it clear that the safety of shipping was still far from adequately secured. Here our information ends. As far as hydrology is concerned, a review (72) of the work done up to 1945 concludes with a mention of the tasks still waiting to be done. One of these is to find out more about the winter hydrological régime in the open sea, knowledge which is needed before it is possible to assess hydrological behaviour over the period of a year. Another difficulty is that some areas have had very much more attention than others, with the result that it is extremely hard to distinguish effectively the general characteristics of any wide expanse of water. It is not known how much has been accomplished under the post-war five-year plan in the sphere of either hydrology or hydrography.

(b) METEOROLOGY

Meteorological records were kept by all polar stations, and by many ships carrying expeditions or parties of scientists. This information was supplemented by the observations of a large number of inland stations which formed part of the regular meteorological network covering the whole of the Soviet Union. In addition use was beginning to be made after the second world war of automatic radio meteorological stations, either on shore or afloat, which functioned unattended (194). The usefulness of all this information was apparent far beyond the confines of the areas with which we are here concerned, since it filled a gap in the world weather map. But we will be considering only its local usefulness.

The need for a weather forecasting service became quickly apparent to Glavsevmorput'. Before 1933 forecasting teams of two men located on board ships had functioned during the summer (see p. 36). In the 1933 season there were six teams of this sort and in 1934 three (44). In the same year the first permanent forecasting centres were set up at Dikson and Mys Shmidta; they were later augmented by three more, at Tiksi, Anadyr' and Amderma. These five stations, which replaced the ship-borne teams, worked all the year round. They collated the reports of the polar stations and ships in their vicinity—the Northern Sea Route was divided into five synoptic zones for the purpose—and broadcast daily weather reports and forecasts (42). Each centre was manned by two meteorologists. Dikson, Tiksi and Mys Shmidta are listed (255) as functioning in 1948. There can be no doubt that these weather reports were a very real help to shipping and aircraft.

While the immediate use of meteorological observations is for forecasting, their equally important long-term use is for determining climatic characteristics. For this observations of a similar pattern over as long a period as possible are required. A number of papers based on the records of polar stations were published; notably a series issued by the Arctic Institute (169), giving detailed climatic studies, by regions, of the whole Soviet Arctic coastline. There have definitely been shortcomings however in the matter of working up results. It was alleged in 1938 that some stations' material had not been worked up for ten years (112). This was no doubt partly due to the absence of a continuous policy on where and how the observations should be published, and this in turn was caused by different authorities having the responsibility for polar stations.

The quality of the observations made was reasonably good. There was a certain wastage of recordings which were unreliable due to inexperienced handling of the instruments. This happened most frequently in the early years. Most of the figures have been accepted by competent judges in the Soviet Union as reliable (169).

(c) ICE AND ICE FORECASTING

The behaviour of sea ice has been the subject of much study. Observations by ships and polar stations on the state of the ice and on factors affecting the ice régime have been collected into annual publications. Analytical papers have been published on sea ice conditions in specific areas, on factors affecting the drift and extent of sea ice, and on its physical and chemical properties (130). The chief importance of this work in relation to navigation is the fact that it helped to make ice forecasting possible.

Ice forecasts for Soviet Arctic waters were first issued (see p. 36) in the 1920's. The object was to provide a general picture of ice conditions in the coming season. The idea of these long-term forecasts, as they are called in Russian, was probably first conceived by Nansen. It was considerably developed by Soviet scientists. In addition to long-term forecasts there arose with the increase of shipping traffic a need for short-term forecasts which would predict in some detail the state of the ice in a given area a few days ahead. This was a later development and became possible only when sufficiently detailed ice reports were available.

V. Yu. Vize and N. N. Zubov were the two Soviet scientists who had been principally interested in long-term forecasting in its early days. After 1932 the obvious importance of the science in its application to planning the shipping season attracted a number of others to its study. Each developed his own method and each produced his own forecast. This was no doubt confusing for the planners, but it was a good way of testing different theories. The Northern Sea Route was divided into its component seas and a forecast was issued for each sea. Forecasts came to be made four times a year: in December, to give a very rough picture of what might be expected the next summer, with the object of helping generally in the arrangements for the winter; in February, to give a more exact version of the December forecast; in May, to provide a basis on which navigation could actually be planned; and in August, to predict the end of the season. In a survey of methods used before 1938, V. S. Nazarov (210), one of the forecasters, makes it clear that almost all his colleagues relied on study of hydrological and meteorological data over the longest possible period for the area with which they were concerned. Some gave special prominence to one particular factor which seemed to them to be of decisive importance: Zubov emphasised the effect of the Nordkap current, a branch of the Gulf-stream which brings warm water into the Barents and Kara Seas; Vize paid particular attention to the "Iceland minimum" of atmospheric pressure, with the movements of which, he suggested, sea ice in the Kara Sea fluctuated; Nazarov stressed the importance of air currents and the density of sea water. One forecaster, A. Burke, evolved a cyclic theory of variations in sea ice. Several attempts have been made to discern such cycles, but none has been successful because past observations of ice conditions are too meagre and inexact to support any theory of that sort. Indeed, it was the lack of data that hampered the development of long-term forecasting as a whole all through the 1930's. Vize (368) complained of it in 1935 and Laktionov (133) was still complaining in 1939. The flights and ice patrols organised to provide the necessary data for short-term forecasts (to be considered later) were not particularly relevant, at least to start with, to the problems of long-term forecasting. Observations from polar stations were only useful to a limited extent. And the observations of the North Pole station of 1937–38, of the *Sedov* drift expedition of 1937–40, and of the expedition to the "Pole of inaccessibility" in 1941, though important, were really only an indication of how valuable regular observations from the central basin would be. However, a definite step forward was taken in 1938 when all work on long-term forecasts was concentrated in the Arctic Institute (101). It introduced an improved scale of measuring the quantity of sea ice; divided the seas into smaller areas for forecasting purposes so that forecasts were less general in application; and took steps to ensure that it obtained all available relevant hydrological and meteorological data from existing stations. In 1940 the Arctic Institute published (323) a list of 23 hydrological sections from the Greenland Sea to Bering Strait; these were to be made at specified times—generally once or twice a season—and were designed to provide the necessary regular information required by long-term forecasters and not available from other sources. A large number of expeditions would have been needed to work the whole programme, but even if it was over-ambitious, its publication shows at least that an attempt was being made to tackle the problem.

While it was possible to have a shot at making a very general long-term forecast on the basis of what was admitted to be too little material, adequate observations were essential before even an attempt could be made to provide a short-term forecast, which implied giving detailed predictions for a comparatively small area. Such forecasts were closely related, clearly, to weather forecasts. The main factors influencing movement of sea ice over a period of days are easier to discern than those which influence it from year to year. The principles of short-term forecasting are therefore not so much in dispute. Karelin (102) outlines them thus. Atmospheric pressure affects sea ice in two ways. First, winds, following the isobars, incline to the left, and ice driven by wind inclines to the right; therefore ice tends to drift roughly along the isobars, with the high pressure on the right. Secondly, if the pressure gradient is steep the ice tends to be compact, while a gentle pressure gradient allows it to drift apart. Sea currents have a big influence, while tidal streams have little except just off shore. And finally, air and sea temperatures cause thawing or freezing. The necessities for short-term forecasting are therefore ice charts, hydrological information, synoptic charts and weather forecasts. Of these, once weather forecasts were available, it was the ice charts which presented the greatest difficulty. Polar stations and reports from ships and aircraft provided an increasing amount of information, but even then there were large areas of sea about which nothing was known. Special measures were therefore taken in order to obtain fuller knowledge of the extent of sea ice. First, increased use was made of aircraft. Since 1924 aircraft had been used for ice reconnaissance, but they were generally attached to a specific convoy and reconnoitred the ice ahead of it. As more aircraft became available, some[1] were allotted to each sea and flew sorties planned to provide the information needed by the forecasters. This system started about 1937. In 1939 the flight plan was extended to cover areas far north of the shipping lanes, and the period of reconnaissance was lengthened. In 1940 the flights took place from April to October, a further lengthening; and from 1941 winter ice reconnaissance flights were regularly undertaken. As a result some of the observations had a bearing on long-term forecasts and were useful in that respect. Secondly, besides the use of aircraft, small vessels were detailed to patrol the southern edge of the consolidated pack ice and report its position, and at the same time to perform hydrological work. Two boats did these patrols in the northern parts of the Barents and Kara Seas in 1938 (52) and 1939 (146), one in 1940 (339), and during the war three or four boats patrolled all the seas (60). Although from the available evidence it does not seem likely that patrols covered anything like the whole ice edge for the whole season, nevertheless their reports were clearly of importance to forecasters.

Whereas long-term forecasts were made by men sitting at desks in Moscow and Leningrad, short-term forecasting required by its nature a different organisation. The first forecasts were made at Dikson in 1937 by B. I. Ivanov, a scientist on the staff of the polar station there, who found that the presence of the weather service group and of the radio centre enabled him to work out

[1] Russian sources do not agree as to the number of aircraft used each year for ice reconnaissance. The list given by A. F. Laktionov (130), covering the period 1929–44, probably gives a rough general idea, but it is frequently contradicted by accounts of individual years.

and transmit this new type of forecast. Ivanov continued the work in 1938. The next year the Arctic Institute (for which he worked) kept him at Dikson with instructions to cover the Kara and Laptev Seas, and sent a team to Mys Shmidta to cover the eastern seas (103). In 1940 three teams worked at Dikson, Tiksi and on board the icebreaker *Kaganovich* which carried the Director of Operations for the eastern sector. Forecasts were for five- or ten-day periods. It was found that the team on the icebreaker was the most useful since its proximity to the staff which controlled shipping movements provided opportunities for informal consultations (383).

There can be no doubt of the importance of both sorts of ice forecasting. It is clear that the period we have reviewed shows both in their infancy, struggling with a lack of essential raw materials. Yet reasonably satisfactory results were produced. It was claimed (101) in 1940 that the correctness of long-term forecasts averaged 75 %,[1] though what period this average covers is not made clear. And short-term forecasts in 1939 showed the following average correctness (104): Kara Sea, 70 %; Laptev Sea, 65 %; East Siberian Sea, 83 %; and Chukchi Sea, 85 %. If these figures are accurate, then the forecasts must have been of some use. But there was obviously room for improvement, and Papanin (233) stressed the point in 1941.

(d) TERRESTRIAL MAGNETISM

In high latitudes the magnetic compass tends to become unreliable as a result of the proximity of the magnetic pole and the frequency of magnetic storms. The necessity for special study of the problem was recognised long before 1933. Under Glavsevmorput', however, work was intensified. The increased number of expeditions in the field resulted in a larger number of determinations of magnetic variation; and permanently functioning magnetic observatories were set up at certain polar stations, as we have seen. Magnetic charts were compiled on the basis of the recordings obtained. According to Gakkel' (61), writing in 1945, the results achieved up to that time were useful to navigators, but the problem was far from being solved.

(iii) *Scientific institutions*

Glavsevmorput' was established in order to provide unified control of shipping on the Northern Sea Route, and everything connected with it. We have already seen (see pp. 34, 36) that co-ordination of scientific work also was a necessity. It was therefore natural that as a step towards securing this Glavsevmorput' should set up within its jurisdiction a principal scientific organ.

The Arctic Institute,[2] which was selected for this position, was the obvious

---

[1] Percentage correctness of forecasts is a very rough guide, calculated on this basis: quite correct, 100 %; partially correct, 50 %; wrong, 0 %.

[2] The Institute was successively known as the Commission for the Study of the North [Komissiya po Izucheniyu Severa], the Northern Scientific-Industrial Expedition [Severnaya Nauchno-Promyslovaya Ekspeditsiya], the Institute for the Study of the North [Institut po Izucheniyu Severa], the All-Union Arctic Institute [Vsesoyuznyy Arkticheskiy Institut], and the Arctic Research Institute of Glavsevmorput' [Arkticheskiy Nauchno-Issledovatel'-skiy Institut Glavsevmorputi].

choice. It had been founded in 1919 and had undertaken on an increasing scale study of natural resources in the far north. During the 1920's expeditions were sent out each year, on land and sea, to the sector between Murmansk and the Yenisey. The icebreaker voyages to Zemlya Frantsa-Iosifa [Franz Josef Land], already mentioned, were organised by the Institute. From 1930 it recognised the whole of the Soviet Arctic as its province, and serious attention was given for the first time to the central and eastern sectors. At the same time the Institute was relieved of responsibility for study of the Murman coast and the Barents Sea. Thus by 1933 the Institute had considerable experience of scientific problems in the Arctic. When Glavsevmorput' took over the Institute the link between the two was already strong, since the first Head of Glavsevmorput', O. Yu. Shmidt, had been Director of the Institute before his appointment.

The Institute's work expanded in relation to the growing authority of Glavsevmorput'. In 1936 the principal departments were for geology; for cartography and geodesy; for hydrology; for industrial biology (hunting and fishing); and for reindeer husbandry. The importance attributed to each department may be judged from the number of staff employed by each for expeditions and research in 1937 (55): the departments of geology and of cartography and geodesy had a combined total of 640; the department of hydrology, 136; the departments of industrial biology and reindeer husbandry, 140. The pronounced leaning in favour of geology was perhaps occasioned by the fact that R. L. Samoylovich, the Director both in the Institute's early days and in succession to Shmidt, was a geologist.

When the constitution of Glavsevmorput' was changed in 1938 and its scope lessened, the Arctic Institute was directly affected. The Institute, like its parent body, was accused of dissipating its energies on non-essentials, to the detriment of problems directly affecting the sea route. Several scientists on the staff, including Samoylovich, lost their jobs and were disgraced. Hydrology, meteorology and study of ice became the main tasks (57). In 1940 the principal departments were those of the ice and weather service, which included a forecasting section; of sea hydrology; and of geophysics (260). The shift in emphasis is further made clear by the fact that the sum of money allowed to the hydrological department in 1939 was six times the 1938 figure. The Institute retained this character during the war, and acquired some small ships which were used for patrolling the ice edge. In 1945 another expansion began: some geological work was reintroduced and new sections were formed for the study of general geography and economics (57). This expansion may continue, for the Second All-Union Congress of Geographers, meeting in 1947, passed a resolution (95) recommending that the Institute be turned into "an institution whose basic task is the complex exploration of the Arctic", and noted its backwardness in the study of botany, zoology, geomorphology, history and general geography.

The Arctic Institute, it is quite clear, did a good job for Glavsevmorput' in the sense that its own scientific work was sound. It planned and carried out a large number of expeditions; it assumed the scientific direction of Glavsevmorput' polar stations; it carried out research on meteorology and on sea ice, and was responsible for ice forecasting; potential sources of fuel were discovered and surveyed. In so far as we are here considering scientific support given to the sea route and to projects directly related to it, the changes of 1938–39 in the structure of the Institute must be regarded as being on the whole beneficial, though we may deplore the atmosphere of political purge which attended them. The greater part of the geological and all the biological work, forbidden in 1938, had in fact been irrelevant to the main issue.

But it is by no means so clear that the Institute was able to bring about, particularly in the early stages, the overall co-ordination that was needed, even within the system of Glavsevmorput'. In 1936 there were complaints (304) that the multiplicity of scientific organisations was creating confusion and duplication of effort, and the example was quoted of four separate expeditions working in the Khatanga estuary in 1934–35: all were organised by Glavsevmorput' but there was no co-ordination between them. Another complaint (113) was heard in 1939 that control over polar stations was divided among three departments of Glavsevmorput' and the Arctic Institute, to the detriment of the work done. But the fault in cases like these lay not so much with the Arctic Institute as with Glavsevmorput', which had appointed the Institute its principal scientific organ and then permitted other departments to do scientific work independently. However, in one field at least steps were taken to end such unsatisfactory situations. In 1938 all work on ice forecasting was concentrated in the Institute, to which two other institutions working on the subject handed over all relevant material in their possession (59). It is not known whether similar co-ordination was introduced in other branches of science; but it is arguable that the reason why the Institute was permitted to expand again after the war was that it had now become an effective co-ordinator of effort.

One important branch of science, hydrography, has been kept deliberately separate. Glavsevmorput' had initially entrusted this work to the Institute. But the quantity of work was found to be so large that very shortly this department was taken from the Institute and became the Hydrographic Administration of Glavsevmorput' (56). This arrangement remains unchanged since 1933, and no other body performs hydrographic work on the Northern Sea Route. The Soviet naval hydrographer, who had been responsible for the work before 1933, appears to rely entirely on the Hydrographic Administration of Glavsevmorput' for information on the northern seas. The Hydrographic Administration maintains branches within its area, and is responsible for both sea and river hydrography. The hydrological department of the Arctic Institute co-operates with it in the production of sailing directions. Although there have been shortcomings in the work done, they have not been the result of an intrinsically faulty organisation, but rather of inefficiency within it. In order to provide qualified men, Glavsevmorput' set up its own training centre in 1935 (253). This was known as the Hydrographic Institute [Gidrograficheskiy Institut] until sometime during the war, when the name was changed to the

Arctic Sea Training School [Vyssheye Morskoye Arkticheskoye Uchilishche]. By 1940 it had produced 65 hydrographers (253).

Scientific work on the Northern Sea Route was not confined to Glavsevmorput' and its organs. The State Hydrological Institute [Gosudarstvennyy Gidrologicheskiy Institut],[1] which was founded in 1919, worked quite frequently in northern waters, and had done some work on ice and ice forecasting. After about 1935 the Hydrological Institute tended to do less work in the north, and in 1938 all ice forecasting responsibilities were handed over to the Arctic Institute. Co-operation between the two Institutes seems to have been good. Several scientists in fact worked for both, notably V. Yu. Vize, who was responsible for the Hydrological Institute's ice forecasts.

From 1929 the State Hydrological Institute was affiliated to the meteorological authority of the Soviet Union. The Meteorological Service, to give it an abbreviated title,[2] had, as might be expected, considerable interest in the Arctic. Its own network of weather stations covered the Arctic and included some in coastal areas. The decree of 17 December 1932 (312) ruled that all existing meteorological and radio stations on the coast and islands in the Arctic should be transferred to Glavsevmorput'. Those belonging to the naval hydrographic organisation and to the Arctic Institute were at once transferred, but the Meteorological Service retained many of its stations for several years. They were gradually taken over by Glavsevmorput', though there is evidence that some marginal stations were still in the hands of the Meteorological Service in 1945. There was, of course, always an arrangement between Glavsevmorput' and the Meteorological Service giving each access to the other's observations, but certain difficulties arose concerning the programme of work, and were thought in 1939 to be having a bad effect (113). These difficulties may have been overcome during the war, when the work of the meteorological stations became of great importance.

The fisheries investigations in the Barents Sea were continued by the State Oceanographical Institute [Gosudarstvennyy Okeanograficheskiy Institut] in Moscow. This Institute was also under the control of the Meteorological Service. Hydrological sections were made at regular intervals in specified places. In 1933 the Institute opened a polar branch known as the Polar Research Institute of Sea Fisheries and Oceanography [Polyarnyy Nauchno-Issledovatel'skiy Institut Morskogo Rybnogo Khozyaystva i Okeanografii], sometimes abbreviated to the N. M. Knipovich Polar Institute [Polyarnyy Institut imeni N. M. Knipovicha] (110); and this branch continued and expanded the same activities, which were however still limited to the Barents Sea and

[1] Known as the Russian Hydrological Institute [Rossiyskiy Gidrologicheskiy Institut] until 1925.

[2] It was called successively the Hydrological and Meteorological Committee of the U.S.S.R. [Gidrometeorologicheskiy Komitet SSSR], the Single Hydrological and Meteorological Committee of the U.S.S.R. and R.S.F.S.R. [Yedinyy Gidrometeorologicheskiy Komitet SSSR i RSFSR], the Central Administration of the Single Hydrological and Meteorological Service of the U.S.S.R. [Tsentral'noye Upravleniye Yedinoy Gidrometeorologicheskoy Sluzhby SSSR], and the Chief Administration of the Hydrological and Meteorological Service of the U.S.S.R. [Glavnoye Upravleniye Gidrometeorologicheskoy Sluzhby SSSR].

occasional excursions into the Kara Sea. As a result of all this work, which had been carried on with only fairly short gaps since the turn of the century, the Barents Sea became the best studied Arctic sea in the world. And in view of the acknowledged effect of the warm currents crossing the Barents Sea on hydrological and ice conditions farther east, this was a factor of importance to the Northern Sea Route.

Two more scientific institutions must be mentioned here, though the contribution of each was somewhat different to that of others already mentioned: the Academy of Sciences [Akademiya Nauk] and the Geographical Society [Geograficheskoye Obshchestvo]. From its foundation in 1725 the Academy was for more than a century the only organised body of scientists in the country, and it was responsible for a considerable amount of work in the Arctic during the eighteenth and nineteenth centuries. But after the Revolution, Arctic work along the coast and in the seas and islands north of the mainland was largely taken over by more specialised institutions. The Academy, which was still the biggest and the senior scientific body in the country, retained an active interest in the north through some of its affiliated institutes, but its work after 1932 affected the Northern Sea Route only indirectly. The main concern of the Academy was with the study of natural resources on the mainland. But the V. A. Obruchev Institute for the Study of Permanently Frozen Soil [Institut Merzlotovedeniya imeni V. A. Obrucheva], which was affiliated to the Academy, conducted research on construction techniques on permanently frozen ground and these had a bearing on problems of port building; and the Murman Biological Station [Murmanskaya Biologicheskaya Stantsiya] (see p. 33), which had ceased to function in 1933, was re-established by the Academy in 1938 and continued to work in the Barents Sea. The most significant influence of the Academy, however, was not by direct contribution of scientific knowledge but by virtue of the fact that some of the most notable scientists working in the Arctic—O. Yu. Shmidt, P. P. Shirshov, A. A. Grigor'yev, V. Yu. Vize and others—were or became its members.

The Geographical Society[1] was founded in 1845 and has played an active part in the exploration of the Arctic. Since the Revolution, however, the Society has tended to undertake less field work on its own account and to become more of a centre and club. It nevertheless remains a lively institution. Its influence has been indirect and is exercised, as in the case of the Academy, through those of its members—and the number is large—engaged in Arctic work.

The foregoing may be summarised thus. The dominating influence in the scientific sphere of which we have been speaking is Glavsevmorput', and its principal scientific organ has apparently been fairly successful, after initial

---

[1] Successively known as the Imperial Russian Geographical Society [Imperatorskoye Russkoye Geograficheskoye Obshchestvo], the Russian Geographical Society [Russkoye Geograficheskoye Obshchestvo], the State Russian Geographical Society [Gosudarstvennoye Russkoye Geograficheskoye Obshchestvo], the State Geographical Society [Gosudarstvennoye Geograficheskoye Obshchestvo], and the All-Union Geographical Society [Vsesoyuznoye Geograficheskoye Obshchestvo).

failure, in co-ordinating scientific effort as between departments of Glavsevmorput'. But the latter does not have a monopoly in the study of science on the Northern Sea Route. Other scientific bodies which had worked there before 1932 continued to do so. There was some overlapping of functions. Geology, with which we are not directly concerned, in fact provided worse examples, particularly when Glavsevmorput' was responsible for development of natural resources on the mainland north of lat. 62° N. Ashby (11) has shown that there are a very large number of scientific institutions in the Soviet Union and that much overlapping does occur. It is particularly likely to happen in the case of studies which are concerned with various branches of science sub divided on a regional basis. The scientific problems posed by the Northern Sea Route have fared better than others from this point of view.

6. USEFULNESS OF THE NORTHERN SEA ROUTE

(i) *Economic usefulness*

Economic motives played a large part in the early development of the Northern Sea Route. They are nowhere more clearly seen than in the history of the Kara Sea route up to 1932. Wiggins, the various trading companies, Lied, and even to some extent Nordenskiöld, were concerned to start a new trade route between the interior of Siberia and western Europe. They encountered difficulties, but their efforts showed quite clearly that when those difficulties were overcome the Kara Sea route could attract enough business to make it very profitable.

The motives which led the Russian Government to open up a sea route from the Pacific to the Kolyma, and later to the Lena, were also economic: a cheaper route than those previously used was sought. But in this case there were no tempting commercial prospects, since imports were confined to supplies for small settlements, and there was nothing worth exporting, as far as was known, from this largely unexplored part of the country.

In attempting to discern the part played by economic considerations after the establishment of Glavsevmorput', we are immediately faced with a difficulty. If we had been considering an enterprise functioning in a free economy, a good pointer would have been whether Glavsevmorput' showed a profit or a loss. But in the Soviet Union, a country with a planned economy, that question has been irrelevant. As Stalin said in 1933 (322): "Profitableness must be judged from the standpoint of the economy of the whole nation and over a period of several years." All we can do therefore is to piece together any relevant information and draw our own conclusions from it.

We may distinguish two functions of the Northern Sea Route. First, as a means of reaching raw materials which would probably not otherwise be exploited—the conception implicit in Stalin's attitude, quoted earlier (see p. 37), that the object of northern development was the incorporation of natural resources into the socialist economy. This function includes not only export of the material, but import of men, stores and equipment necessary to get out the material. Second, as a passenger and freight route in competition with other routes; for instance, it may be quicker, or cheaper, or more convenient to send goods from Moscow to Vladivostok by the Northern Sea Route rather than by any other route. These two functions may overlap at times, but it will make for clarity if we keep them separate as far as possible.

When we consider the Northern Sea Route as a means of making raw materials accessible, timber is the material which first comes to mind. Because of its large bulk it is transportable much more economically by water than by land. It grows in very large quantities in the vast *tayga* (virgin coniferous forest) zone, which stretches across almost the whole width of the country and whose northern limit varies between lat. 66° and 70° N. Excellent quality softwoods are available. The timber of the lower Yenisey was already in 1932 virtually the only export by the Kara Sea route (see p. 21). The volume continued to rise up to 1939, which is the last year for which figures are available.[1] It is true the first five-year plan envisaged a very much larger increase in traffic: turnover in 1933 was to be 796,990 tons (244) carried in 300 ships, while in fact about a seventh of that total was carried in thirty ships. But although the discrepancy between plan and accomplishment was so great, for the period 1933–39 the Kara operations—which means in effect the timber industry based on Igarka—accounted each year for between 50% and 75% of the total freight turnover of the Northern Sea Route. The purpose of the industry continued to be export to western Europe. This is made clear by the fact that Glavsevmorput' relinquished control of the business side of the Kara operations in 1938 and Narkomvneshtorg [the People's Commissariat for Foreign Trade] took it over, in collaboration with Sovfrakht, the state shipping firm. An important point arises here. There would be little reason for the existence of the industry if it were working only for the home market, since timber is abundantly available in many less remote parts of the Soviet Union. There would be no point at all in taking it upstream from Igarka to the industrial areas since the latter can get all the timber they want from the southern *tayga* areas. Nor would there be any point in shipping it across the Kara Sea to Murmansk since that area is equally well supplied locally. Therefore this application of the sea route, accounting for such a large percentage of the total turnover, is entirely dependent upon foreign trade. The post-war state of the industry is not known. But these facts seem significant: very few British or Norwegian ships have been chartered for the Kara Sea voyage since the war; and of the 1949 purchases of Soviet softwoods by the United Kingdom only 16,983 standards out of a total of 124,484 were to come across the Kara Sea (and in fact only 9683 could be shipped).[2]

The Ob' basin also contains vast quantities of timber, and Glavsevmorput' was building sawmills at Belogor'ye, near the Ob'-Irtysh confluence, in 1936. But there is no record of any sizeable timber export by sea; presumably it was ruled out by the absence of a port comparable to Igarka.

The Lena also flows through heavily wooded country, but there were

[1] See Appendix I.
[2] Information on 1949 purchases was supplied by the Board of Trade Timber Control, May 1950.

practical difficulties in the way of exporting timber by sea. Sea-going ships could not enter the river, and the timber could not be rafted to Tiksi because of the danger that the rafts would break up in the final stretch of open sea; therefore barges had to be used, and this limited the quantity that could be taken downstream. In 1939 however 10,000 cu. m. of timber were rafted down for the first time, evidently as a result of improved rafting technique (254). In 1940 timber was shipped from Tiksi to Bukhta Provideniya (223). There is not enough evidence to show what use was intended to be made of the Lena timber. The intention may have been to supply the local needs of Chukotka, which is not well supplied with standing timber; but these needs are not great. To the south, the more populous parts of the Pacific seaboard have plenty of local wood. Possibly it was later intended to export Lena timber to western Europe, since the larch (*Laria sibirica*), a very high-quality wood, grows in much greater quantities in Yakutiya than on the Yenisey; or conceivably the object was export to Japan or China, both of which took considerable quantities of Siberian timber in the early 1930's. Whatever the intentions, however, it may be said that the practical possibility of getting timber out of the Lena by sea has been demonstrated.

After timber, minerals are the most promising freight. Coal and oil will not be mentioned here since it has already been made clear that any locally mined fuels of reasonable quality will be needed for domestic use in the ships of Glavsevmorput'. These fuels cannot therefore be included among the raw materials that the Northern Sea Route has made available to the country until a surplus is available for other uses. But the local use of Arctic-mined coal is important in so far as it relieves the coal-producing centres in the rest of the country by rendering the ships of Glavsevmorput' and local Arctic industrial enterprises self-supporting in this respect. The fact therefore that both these needs appear to be well on the way to being met from Arctic sources does increase the value of the Northern Sea Route in other respects.

The presence of salt in the Nordvik area was noticed by Khariton Laptev in the first half of the eighteenth century, but it was not until 1933 that serious attempts were made to study the possibility of mining here. Work was slow and intermittent. Some 2000 tons of salt were quarried by hand in 1936, and a small proportion of this was sent by sea to the Pacific. Later a deposit of higher quality was found, and mining started in 1942. In the three years 1942–44, 38,000 tons were shipped out to the east (139). The enterprise required no coal from outside, since the local coal is sufficient for this purpose. The importance of the salt lay in the fact that Far Eastern fisheries were being developed, and the salt necessary for preserving the fish—calculated at 200,000 tons for the 1940 season—was not available anywhere in the region of the Pacific seaboard, with the result that long train hauls over the Trans-Siberian railway were unavoidable. There was therefore an assured market for Nordvik salt; but a report (109 a) on salt used by the Far Eastern fisheries from 1945 to 1950 does not list Nordvik among the important producing areas.

The Soviet Union has been very short of tin, which before the second world war had largely to be imported. Intensive efforts were made to locate and work tin deposits, and a number were found, principally in the interior of Siberia. It was claimed(23) after the war that the newly found deposits would soon render imports of tin unnecessary. There has clearly therefore been a big increase in production fairly recently; and this fact lends weight to the allegation, made by one writer(63), that tin-mining started in the area of Chaunskaya Guba during the war and that output exceeded that of all other tin-mines in the country. Unfortunately there is no confirmation from a second Soviet source that production has started. But there is geological evidence that tin is found here(188); and since 1940 a port of some sort has been built at Pevek, which is the site of one of the tin-bearing localities, and between five and eight ships carrying principally lend-lease food and oil were to call there during each of the war years. If mining is in fact going on, the Northern Sea Route is essential to the project.

Several mineral deposits in the Tungus basin (east of the lower and middle Yenisey) have been worked. It will be remembered that graphite from the Kureyka was exported through the Kara Sea in the nineteenth century. Other deposits were later found in the same general area. But carriage of graphite by ships of the Kara operations tailed off as the industrial development of the interior of Siberia proceeded, and in the 1930's graphite from the Tungus basin was taken upstream to Krasnoyarsk for refinement. There is a more recently founded mining centre at Noril'sk, near Dudinka on the lower Yenisey. We have spoken earlier of the coal deposits here. From the middle 1930's nickel, and probably other metals, have also been mined. Unfortunately no production figures are available, nor is there any information as to whether the Kara Sea route normally carried the output. While this source of nickel is admitted to be useful, it is not the only nor the best source in the country.

Timber, salt, probably tin and possibly nickel—these are apparently the raw materials which the Northern Sea Route had, until recently, been able to make accessible. Two other mines which are known to have been in production —lead on Ostrov Vaygach and fluorspar at Amderma—are not included because they stand right at the western entrance of the Northern Sea Route and cannot therefore be said to be made accessible by it. Let us examine the achievements of the route in its other function, as a competitor with other freight routes.

First let us consider the route as a link between European Russia and Far Eastern Russia. Omitting air and road transport, which can only be used for long-distance freight in special circumstances, there are four routes connecting Moscow or Leningrad with the Far East: the Trans-Siberian railway, by sea from the Black Sea via the Suez Canal, the same from the Baltic via the Mediterranean, and the Northern Sea Route. The railway is the shortest distance, and is of course much quicker than any of the sea routes; but it is more expensive, and is normally so overworked that efforts are made to divert traffic into other channels. Of the three sea routes, the Northern Sea Route is substantially the shortest: Vladivostok is 7000 sea miles from Arkhangel'sk via Bering Strait, while it is 10,800 sea miles from Odessa and 14,700 sea miles from Leningrad via the Suez Canal. On the other hand the southern routes can be used all the year round while the northern one is limited to about

PLATE V

Part of the mining settlement at Noril'sk, photographed in 1935. In foreground, the narrow-gauge railway which later linked Noril'sk with Dudinka on the Yenisey.

PLATE VI

Tugs and barges of the Lena river fleet lying up for the winter in berths cut out in the ice in a sheltered backwater. Here they escape damage when the river ice breaks up in the spring. In the foreground, the *Partizan Shchetinkin*, built in Germany and acquired by the U.S.S.R. in 1928.

2½ months. However, there is on the face of it every chance that during its short season the Northern Sea Route may be both quicker and cheaper than the southern routes, and cheaper than the railway. In practice the railway was found in the early stages to be cheaper. In 1935 transport of a ton of grain from Leningrad to Vladivostok cost 275 roubles, which was 50 % greater than the rail charge; the 275 roubles included 135 for "special expenses"—services by icebreakers, aircraft and polar stations. This information is provided by

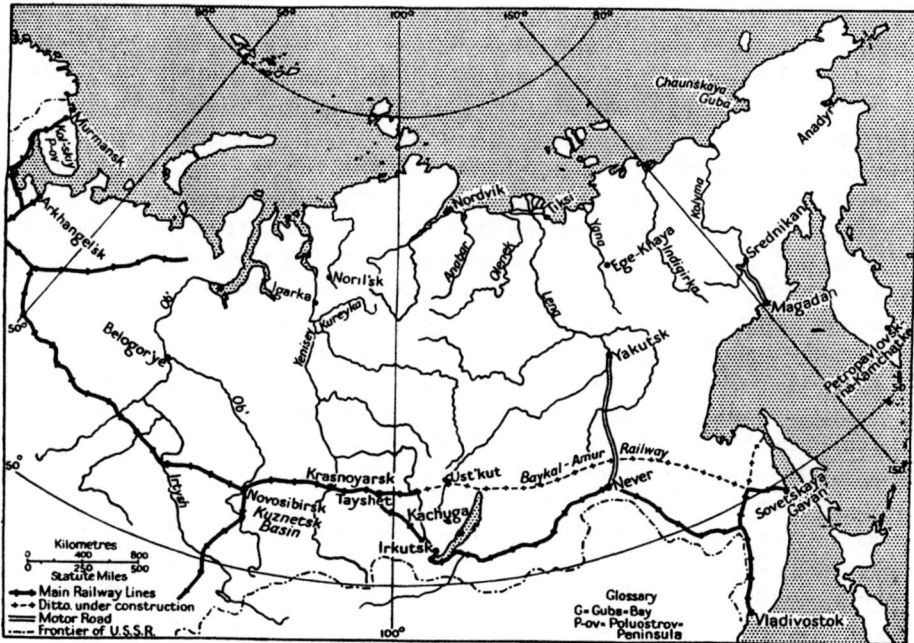

Map 7. Other communication systems affecting the economic
usefulness of the Northern Sea Route

S. Ioffe (83),[1] at one time Deputy Head of Glavsevmorput'. O. Yu. Shmidt (289) however states that in 1937 the Northern Sea Route was already cheaper than the railway. Both Shmidt and Ioffe agreed, as one would expect, that the charges would shortly go down considerably as traffic and efficiency increased. Unfortunately no further figures are available to show whether the charges in fact went down. We have seen that the total turnover tended to increase up to 1940. On the other hand the number of through voyages appears not to have increased—rather the contrary; and icebreaker and aircraft services grew larger. It is therefore difficult to come to any conclusion. It seems likely that the margin between rail and sea freight rates will not have been great, and that

[1] The same figure of 275 roubles is also given by B. V. Lavrov and N. Ye. Shadrin (142), but it is said to be twice the rail charge and to include no "special expenses". Ioffe's version is preferred in view of the apparent unreliability of the other paper, pointed out particularly by S. P. Natsarenus (205).

the Northern Sea Route will have been able to compete economically with the railway only in the carriage of freight coming from and going to points close to the terminal ports; and as far as can be judged such traffic is not large. It is possible, however, that the Northern Sea Route will have acquired a decided advantage in serving places on the northern part of the Pacific seaboard, far from Vladivostok—places such as Anadyr and Petropavlovsk-na-Kamchatke.

The use of the Ob' and Yenisey as links with the interior of Siberia seems to have suffered eclipse since the creation of the timber industry on the lower Yenisey in the late 1920's. The freight turnover figures for the Kara Sea traffic up to 1939 show that timber shipped at Igarka virtually replaced all other freight from the Yenisey, and the Ob' fell into disuse doubtless because of the unsatisfactory situation of the port at its mouth. There had been plans for making fuller use of both rivers. Ioffe (84) suggested in 1936 that grain should be carried from western Siberia to Kol'skiy Poluostrov, where mining industries were growing; and that apatite mined there should be taken back to be used in the preparation of fertiliser. Presumably the reason the grain was not shipped was that it was required locally together with all other agricultural produce to support the new industrial areas of the Kuznetsk basin. The apatite was also apparently not shipped, for imports to Igarka dropped to only a few thousand tons. No explanation for this is forthcoming. Apatite would have been an ideal cargo which could easily have been loaded in empty ships going to Igarka. Failure to use these ships seems extraordinary; and there were complaints in 1939 that they were not even used to carry stores to places in the immediate vicinity of Igarka, such as Dudinka. On the face of it there seems to have been inefficiency and confusion here.

It is true however that river craft have been brought to the Ob' and Yenisey systems by this route. We have seen that this was the case in 1949 (see p. 50). The first voyage from Leningrad to Krasnoyarsk by way of the White Sea canal, the Kara Sea and the Yenisey was made by a motor-boat in 1938 (372).

The Lena fared better than the Ob' and Yenisey. It was always assumed that the Northern Sea Route would become the principal carrier of heavy freight to central and northern Yakutiya. Before the sea route existed the normal route to Yakutiya was by the Trans-Siberian railway to Irkutsk; by road to Kachuga on the upper Lena; by raft down the river to Ust'-Kut, where river craft carried on downstream to Yakutsk. From Moscow to Yakutsk by this route was 4809 miles. By the Northern Sea Route the distance was only slightly less—4655 miles—but the route was expected to be cheaper and also, surprisingly, quicker. This was in fact the case. In 1935 the sea freight rates were roughly 5 % cheaper than the inland rates; and goods reached destinations in Yakutiya in some six to seven months, as opposed to an average of ten months overland (145). In 1939 Papanin (235) declared that the sea route rates were 47 % cheaper than the inland rates. It seems clear that the Northern Sea Route was fulfilling a useful purpose here. The traffic consisted entirely of supplies for the inhabitants of Yakutiya. Apart from the timber exports which were just starting in 1939, and the coal from Sangar-Khaya which was used for bunkers at Tiksi and Bukhta Provideniya, nothing left

Yakutiya by sea. This was largely because the only known exportable products of the area came from the gold and other mines situated in southern Yakutiya where the distance from the railway was not excessive.

Recent developments in inland transport however may have had some effect on the sea route to Yakutiya. It is likely that in recent years a railway has been built from Tayshet, on the Trans-Siberian line, to Ust'-Kut on the Lena. This would form part of the Baykal-Amur railway which is to run eastwards to the Pacific, north of the Trans-Siberian and parallel to it. This line is known to have been under construction since before the war, but progress has not been made public. If Ust'-Kut is on the railway, then the road and raft section of the inland route to Yakutsk would be eliminated. This would undoubtedly lower costs and increase speed of transport considerably. Another potential rival of the sea route is a road, which has definitely been completed, from Never on the Trans-Siberian through the Aldan gold-mining district to Yakutsk. This road will not be cheaper than the sea route, but it will probably be open all the year round. The effect of these two developments is hard to judge, but it is likely to be appreciable.

If there were advantages in a sea route to the Lena, there were still greater advantages in a sea route to the other rivers of Yakutiya. Supplies for settlements on the Yana, Olenek and Indigirka were formerly sent on overland from Yakutsk or other points on the Lena. This journey was entirely avoided if the supplies came straight up the rivers. The volume of traffic however was very small, certainly up to about 1938. But there may have been important developments since then. One of the recently found tin deposits—the largest in the U.S.S.R.—is at Ege-Khaya, near Verkhoyansk on the upper Yana, and work is alleged (14) to have started here in about 1938. No information on the communications used by these mines has been traced. A road may connect them with Yakutsk, and thence the Trans-Siberian railway. But there is every likelihood that the sea route will have been used in any case for heavy goods.

The Kolyma is another link between the Northern Sea Route and industrial areas to the south. The industry in this case is gold-mining, which started in earnest on the upper reaches of the river in 1932 under the control of Dal'stroy. As a result the Northern Sea Route at first enjoyed a boom; large freight consignments were taken from the Pacific to the mouth of the Kolyma. But this was only temporary, for Dal'stroy decided to make its main line of communication the Okhotsk Sea. In Tsarist times the Kolyma district had been supplied in this way, but it was found that the overland transport from the north coast of the Okhotsk Sea to the upper reaches of the river was expensive. Dal'stroy however was a large organisation disposing of a big force of convict labour, so it was able to build a motor road from Magadan, which was selected as the main seaport, to Srednikan on the Kolyma. Magadan is ice-free for six months in the year, while the mouth of the Kolyma may be accessible for only two. Once this road was functioning there was no need to use the Northern Sea Route for transport to and from the Pacific railhead at Vladivostok or later Sovetskaya Gavan', except possibly for certain exceptional freights. But

Magadan depended entirely upon the railhead for supplies, many of which came from European Russia in the first place. Clearly there was a case here for using the Northern Sea Route. The route Moscow-Murmansk-Kolyma is very much shorter and cheaper than Moscow-Vladivostok-Magadan. In 1935 the first ship sailed from the west to the Kolyma and back. Similar voyages were made in subsequent years. The controlling factor is the ability of the ships to get to the Kolyma and back in one season, rather than any doubt as to the economic value of the route. Besides the Dal'stroy mines there is no other industry or undertaking in the Kolyma valley of sufficient significance to call for heavy transport. The coal-mine at Zyryanka only supplies ports on the Northern Sea Route. There remains the problem of maintenance of local settlements. This is less important than in the case of the Lena because the population is smaller. It is reasonable to assume that the same principles apply here as for Dal'stroy, and that the road is the chief supply line.

It is clear then that the hope that the Northern Sea Route would provide an important alternative route to the interior of the continent has, judging by the information available, been only partially fulfilled. The Ob' and Yenisey are scarcely used at all; the Kolyma is used for the transit of supplies for Dal'stroy from the west, but not from the east; only the route up the Lena and the lesser rivers on either side of it appears to have been accepted for the supply of northern and central Yakutiya, and recent improvements in inland road and rail transport may affect the usefulness of this route. It is significant that almost all the lend-lease goods sent to north Siberian ports during the war were maintenance stores for Arctic undertakings, and were not sent upstream to the interior of the country; presumably it was found more satisfactory to send goods to the Siberian industrial centres by way of Vladivostok, Murmansk or the Persian Gulf. Perhaps one should not exaggerate the economic significance of this, since in wartime other than economic factors may be supposed to have played a part; but it is a pointer. It is not possible to estimate the success or failure of the route as a link between European Russia and the Pacific seaboard. Although there were fluctuations in the number of ships making through trips, traffic did not increase or decrease to a startling degree. Since there is no information on the type of freight carried, it would be premature to come to any conclusion about the economic value of this use of the route.

Of the two economic functions we have been considering it is clear that the first—tapping of otherwise unreachable raw materials—was much the more important. The freight carried, including stores brought in to the undertakings as well as their output, cannot have averaged less than two-thirds of the total turnover of the whole Northern Sea Route in the period up to 1941. This much is clear from the figures for timber shipments from Igarka (given in Appendix I). The other function accounts for much less than the remaining third, since that third must include also maintenance stores for expeditions, bases, ports and polar stations. This order of importance may be interpreted as a healthy sign. As we saw earlier in the case of Kara Sea traffic, the carriage of northern resources provides the firmest basis for operating the Northern

Sea Route; the usefulness of the sea route as a link with the interior is always threatened by possible future improvement in inland transport.

It must of course be borne in mind that the information here given only refers to the pre-war period, and in some cases to wartime as well. There is almost no information on post-war developments. It is worth while to make a few remarks on possible recent trends.

It seems likely that more mineral deposits will have been developed. A list published by the Arctic Institute in 1934 (286) shows 273 Arctic deposits of twenty different minerals, including graphite, gold, sulphites, copper and iron. This list gives no indication of the quality, quantity or accessibility of the ores, and is clearly based in some instances on slender evidence. It would be quite wrong to suppose that anything like all the deposits mentioned would be worth working. But nevertheless it remains true that there is great mineral wealth in the Soviet Arctic—the list of 1934 continued to grow—and the introduction of river craft on to many hitherto unnavigated rivers has without question greatly extended the area in which mining is a possibility. Wartime necessity may have hastened its development.

Fish is another likely freight. The lower Ob' and Yenisey and the two estuaries abound with fish, notably various species of whitefish (the genus *Coregonus*), and sturgeon. These waters have been fished for some time; by the 1870's there were already a number of fishermen who went down to the mouths of the rivers each summer and caught fish by netting. Under Soviet rule canneries and improved preserving methods have been introduced, and as far as can be gathered from more than usually contradictory statistics, catches have increased. The fish is normally taken back upstream and marketed inland. But there can be no doubt that the fisheries could be greatly enlarged. A writer (40) on Yenisey fisheries emphasised in 1941 that the introduction of powered fishing vessels and better tackle, and also the use of hitherto neglected species of fish, could make a big difference. And if in fact a large increase was brought about, then perhaps the sea route would carry some of the surplus to eastern or western terminal ports. It should be noted in this connection that it is precisely at the two ends of the sea route—in the Barents Sea and the Bering Sea—that fisheries have been most developed in Soviet times; but this does not mean that these fisheries satisfy the whole needs of the areas served by their local ports—after all, Murmansk serves much of the northern part of European Russia. It is safe to say that there is still a large market for fish in the Soviet Union. The Ob' and Yenisey fisheries then provide potential freight for the Northern Sea Route. Fisheries farther to the east are still less developed. Preliminary surveys have shown that there are plenty of fish on the lower reaches and in the estuaries of the east Siberian rivers: whitefish (*Coregonus*) again predominate, and there is evidence (135) that the chum salmon (*Oncorhynchus keta* Walbaum), the most valuable fish of the Bering Sea fisheries, is now found in east Siberian rivers.

Sea-mammal hunting may also become important. It has long been practised in the White Sea and Barents Sea, and in the western part of the Kara Sea. Seals are found in all the seas north of Siberia, and walrus in the

Laptev Sea and the Chukchi Sea particularly. Presumably hunting could be extended to the new areas.

The fact that the sea route to the interior of Siberia does not appear to have amounted to very much at the time of the outbreak of war does not necessarily mean that it has no prospects. There was plenty of room for improvement in organisation. Until 1939, for instance, no ships, apart from those on through voyages, left northern waters through Bering Strait in cargo. The first cargoes to leave north Siberia for the Pacific were salt shipped from Nordvik to the Far Eastern fisheries, and coal from Zyryanka to Bukhta Provideniya. Clearly if return cargoes had been found for ships going to the Lena and Kolyma, or servicing polar stations, freight rates could have been lowered. Again, an economic balance does not seem to have been struck between two methods of delivery at small ports: first, large ships stop only at two or three of the principal ports, from which onward transport is provided by coasting steamers or schooners, with the result that the freight charge for the voyage from Murmansk to Tiksi (2050 sea miles) was said in 1939 to be actually less than that for the voyage from Tiksi to the mouth of the Indigirka (510 sea miles) (167). Secondly, it is recorded that a ship carrying 600 tons of supplies to posts in Novaya Zemlya had to make 22 calls, fourteen of them at places which received less than thirty tons; as a result well over half the total charter period was spent at anchor (114). If these sorts of difficulties are overcome, and they certainly do not look insurmountable, the resulting lower freight charges may have some effect on this application of the sea route. It should also be remembered that thousands of miles of useful and potentially navigable waterway—tributaries of the three large rivers and the main stream of several of the lesser ones—were in 1940 either not yet in use or were only just beginning to be used.

There can be no doubt at all that the economic results of the working of the Northern Sea Route during the first eight years (1933–40) were so small as to be out of all proportion to the effort and money expended. But this does not necessarily mean that the economic aims of the whole project can be disregarded and considered a mere cover for something else. The Northern Sea Route, according to Stalin, was brought into being in order to incorporate the wealth of the north in the socialist economy; it was designed, in fact, to meet a general rather than a specific need. The precedence given to construction of ports like Dikson and Bukhta Provideniya, which serve as bases for shipping and are not intended to handle imports and exports, seems to bear this out. We also know that in the planned economy of the U.S.S.R. it was not necessary to show a profit over a short term. Therefore it can be argued that impressive results were not to be expected in this period; that the undertakings served by the sea route up to 1940 were in the nature of pilot projects; and that it is only after the lines of communication are firmly established that conditions are ripe for large-scale development. This argument is supported by the fact that from 1939 there was a growing awareness that the time had now come to pay attention to costs and their relation to results. It is true that for some time certain departments of Glavsevmorput' had worked on a basis of cost accounting

[khozraschet] (see pp. 55, 57), though many of these had been industrial undertakings on shore, over which control was lost in 1938. But in the 1939 season,
called at the time "the first season of trial commercial exploitation", the
system of cost accounting was introduced for small units (242); individual ships,
for instance, had to keep a profit and loss account of the season's work. This
should not be interpreted as a sign that Glavsevmorput' was beginning to pay
its way; the intention, as is clear from an account (17) of the scheme in operation, was simply to make the lower-grade administrators and the rank and
file aware of the importance of not wasting money. The next step was taken
in 1941, when Papanin (234) said that the fleet was to attempt to do without the
Government subsidy which had covered all losses previously. The distance to
go before this objective could be reached may be judged from the fact that the
decree of January 1941 (309) laid it down that 90 % of the employees of the
Moscow office of Glavsevmorput' were to be paid out of the state subsidy, and
only some of the remainder were to be included as a charge on cost accounting.
However, this renunciation of state support was taken by a number of Soviet
industrial undertakings at various times; it might almost be called normal
procedure for any construction project emerging from the pioneer stage.
Unfortunately the war intervened and we do not know what the result was
in the case of Glavsevmorput'. But the fact that the step was taken may
certainly be regarded as a sign that attempts were being made to make ends
meet. It is worth mentioning that a fierce insistence on the economic aspect
was often voiced in official pronouncements by Glavsevmorput'. In an editorial
in the periodical *Sovetskaya Arktika* in 1935 (241) people who suggested that there
might be less expensive alternatives to the Northern Sea Route are accused
of thinking that Bolshevik determination to win through can be shaken by
mere climatic difficulties. B. V. Lavrov, sometime head of Komseveroput'
and of the Institute of Economics of the North [Institut Ekonomiki Severa],
was accused of being a right deviationist when he said in 1937 that the economic expediency of the through route was open to doubt (401).

(ii) *Strategic usefulness*

By using the Northern Sea Route Russian ships may pass between
European and Far Eastern Russia without crossing foreign and potentially
hostile waters and indeed practically without losing sight of the Russian
coast. In addition they are almost invulnerable to attack by hostile ships,
since an enemy would find it very difficult to reach the area without the help
of the Russian weather and ice reports. It was in this that the principal
strategic usefulness of the route was considered to lie. This conception was in
the minds of those who planned the work of the *Taymyr* and the *Vaygach* in
1910–15 (see p. 15); the impetus had been provided by the Russo-Japanese
war, which had illustrated how useful such a route would have been in
allowing the Russian Baltic squadron to reach the Pacific. This was the only
occasion before 1917 on which strategic motives can be said to have played an
active part in the development of the Northern Sea Route. The results of the
expedition were important, but were not a decisive factor. The economic

reasons underlying the early development of the Kara Sea and Kolyma traffic were much more significant.

It is more difficult to discern the part played by strategic considerations after the Revolution. Not only is the Soviet Union reticent about strategic matters, as might be expected; but it is impossible to be certain how far, under the Soviet system, economic considerations may still be effective even though heavy financial losses are being incurred. We have just seen in the last section that heavy losses do not necessarily indicate that the state is not motivated by an economic impulse. Although they do not necessarily indicate it, however, they may do so; and the most obvious alternative to economic motives are strategic motives. But before 1939 there is little actual evidence that strategic reasons influenced the development of the east-west link provided by the Northern Sea Route. It may be possible to read some significance into the fact that when Shmidt made his report to the Government (292) on the results of the 1935 season, the report was addressed to the Commissariat of Defence as well as to the two normal addressees for such communications, the Central Committee of the Communist Party and the Council of People's Commissars. But this may equally be a matter of pure routine. The first open reference to the strategic value of the route was made in 1939, when the shadow of war lay across Europe and everyone was thinking in strategic terms. At the Eighteenth Congress of the Soviet Communist Party in March of that year, Papanin mentioned how the strategic significance of the route had been discussed after the Russo-Japanese war and the disaster of Tsushima, and went on: "Tsushima will never be repeated. And, if need be, our naval squadrons will pass along the Northern Sea Route, will pass along it in order to annihilate the enemy in his territory, on his land and in his waters" (236). He was thinking, of course, in terms of a war with Japan. At the same Congress V. M. Molotov defined the objective of Glavsevmorput' under the third five-year plan as "to turn the Northern Sea Route into a normally working waterway, securing a regular link with the Far East" (342). Though earlier directives had spoken of mastering the Northern Sea Route, this was the first occasion on which this reason—the link with the Far East—was given. A little later we find the commentator Slavin (303) developing this theme that the third five-year plan for Glavsevmorput' was to serve two purposes: defence, and the development of natural resources.

Precisely what strategic use was made of the through route during the second world war is not known. It is likely that naval vessels reinforcing the Soviet fleet based on Vladivostok reached the Pacific by way of the Northern Sea Route. Certainly such an operation would present no difficulty. It is also likely that freighters were transferred as occasion demanded between Murmansk or Arkhangel'sk and Vladivostok. But since the Soviet Union was not, with the exception of six days in 1945, involved in a war in the Pacific, the strategic usefulness of the through route was not really put to the test.

There are other strategic uses for the Northern Sea Route, besides that just mentioned. Access by sea to points on the north coast of Siberia implies several possible strategic advantages: minerals of strategic importance may be reached and mined; thanks to the river system, ports able to serve the interior of the country are available in case other ports are blockaded; military and air bases, strongpoints and rocket-launching sites can be built and maintained. There is no indication in published material that any of these acted as motives for the actions of Glavsevmorput' in time of peace. Possibly the search for minerals which were scarce in the Soviet Union could be considered as dictated by strategic demands; but equally it could have been caused by a desire to be economically independent of foreign markets and their price fluctuations. Two of the strategic advantages mentioned above were probably not made use of even in the war: the evidence provided by the manifests of lend-lease ships tends to show that north Siberian ports were only used for serving local Arctic undertakings; and it was not necessary in a war against Germany to build bases in the far north of Siberia. Probably it was only the access to minerals which was useful to the Soviet war effort.

As far as the available evidence allows us to judge, it seems that strategy was not an important motive in the development of the Northern Sea Route under Glavsevmorput' up to 1939. Strategic advantages did become apparent, but there was little need to make use of them. They are still at the disposal of the Soviet Union. The secrecy which continued to surround activities on the route after the end of the second world war may be taken as probable evidence of recent strategic interest in the area. That such secrecy really is the policy of the Soviet Government and is not just an impression gained by western students of the Soviet Union is proved by a decree of June 1947 (238) which laid down a very wide range of subjects, including aspects of transport, henceforth to be considered state secrets. A proper estimate of the current strategic significance of the Soviet Arctic could only be made after recourse to material not accessible to the public. It is only possible to say therefore that in a hypothetical conflict with America the ability to put up military installations in the far north may assume an importance it has not yet had. On the other hand, the through route may not be an outstanding asset. In spite of its obvious advantages, already mentioned, its short season is still a very severe limitation. There is some advantage clearly in being able to bring up reserve vessels once a year; but it would not be possible to rush a striking force from one ocean to the other to meet a sudden need, unless the need happened to arise between the end of July and the beginning of October.

(iii) *Other uses*

Economics and strategy are two obvious ends which the Northern Sea Route has been made to serve. They are the only two which could have provided a motive for its development, since no other would justify the enormous capital outlay. Other uses however have been made of the route, and these should be mentioned.

Ever since it came to power the Soviet Government has paid close attention to the treatment of national minorities. Soviet communists attribute great importance to the theory and practice of their nationalities policy, since they consider it makes such a flattering contrast with the policy of the previous régime. Although the northern peoples, and especially those living in the tundra areas near the coast, are numerically very small, the principle at stake demanded that every effort should be made to bring home to them the advantages of living in a socialist society. Here the Northern Sea Route was clearly able to play some part. It was used to convey equipment for the setting up of hospitals, schools and community centres, and so-called "cultural goods"—food, musical instruments, household utensils and so forth. This freight totalled only a few thousand tons a year—certainly never more than $2\frac{1}{2}\%$ of the total turnover—and it can only expand appreciably with the growth of the native population. But its importance is greater than its bulk seems to indicate, because of its political implications. Although it has no direct bearing on the exploitation of natural resources, this provision of goods for the natives is really regarded by the Soviet theorists as part of the whole plan for bringing the north fully into the Soviet system.

The Northern Sea Route has also been used as a source of propaganda. Wide publicity has been given by the Soviet information services to activities in the Arctic. The general line taken is exemplified in the message sent by Stalin to the crew of the *Sibiryakov* after its traverse of the whole route in 1932: "There are no fortresses which Bolshevik daring and organisation are not able to storm" (375). Examples of Bolshevik courage, endurance, skill and every other good quality were found among the "conquerors of the Arctic" and were turned to good account by publicists. But clearly the publicity value of the Arctic cannot have been a factor which might have influenced the Government in deciding to embark on a large and expensive development project; there would have been much cheaper ways of securing good adventure stories with a communist moral. Publicity was a useful by-product: useful to the Government for political reasons—to enjoy the results of what Tara-couzio (331) calls "the psychological effect of the spectacular"; useful to Glavsevmorput' as a recruiting device—it is said (12) that the ambition of many Soviet young people now is to become an Arctic flier or scientist.

The cause of science has benefited from the successful exploitation of the Northern Sea Route. Almost every ship that passes along any part of the route could collect some information of value to the scientist, and thanks to the close connection between scientific and administrative echelons this is generally done. The problem of transporting expeditions to and from their area of operations has also been greatly simplified.

There are thus several lesser advantages which arise from ability to use the Northern Sea Route, and they must be taken into account in the final assessment of the route's usefulness.

CONCLUSION

It is undeniable that under Glavsevmorput' the Northern Sea Route was transformed. The evidence we have examined—the traffic carried by the route, the measures taken to equip it, and the scientific support enlisted—make this abundantly clear. No important factor has been neglected, though some features of the advance, such as the handling of icebreakers, have been especially successful while others, such as the development of ports, have been less so. Seen in historical perspective, two points require emphasis. The first is that the idea of trading in these waters was by no means new when this intensive development took place; in fact—and this point is seldom stressed in the Soviet Union—firm foundations for the development had been laid before the Revolution of 1917. The routes through the Kara Sea and to the Kolyma had both been shown to be quite practicable; and the establishment of the first polar stations and the large-scale performance of hydrographic work by the *Taymyr* and *Vaygach* expeditions are evidence of the serious attention paid by the Government to the problem of acquiring the necessary scientific knowledge of the area. The second point is that once the Soviet Government had decided to launch its scheme for accelerated development, results were achieved which could not have been achieved, or at least not so quickly, under capitalism. The history of the Kara Sea route from 1876 to 1899 is one of sporadic attempts to make money, each ending in failure caused by the unwillingness of shareholders to take heavy risks. Lied, the last representative of private enterprise, was successful in overcoming the difficulties; but he found one setback hard enough to deal with, and a run of bad luck would undoubtedly have put him also out of business. The Soviet success is due very largely to the fact that this handicap was removed; the financial security enjoyed by Glavsevmorput' enabled the whole project to forge ahead in spite of major setbacks like the loss of the *Chelyuskin* and the disaster of 1937. Besides the question of finance, however, it is true that other Soviet methods played a part. In particular, the communist belief in the cardinal importance of science led to the performance of much useful scientific work. The remaining methods peculiar to the Soviet Union—the drive to fulfil the plan, stakhanovism, the conferment of decorations and special privileges, the glamorising of the Arctic—may be regarded as substitutes for the profit motive which capitalism would have exploited probably with equal success.

Much has been achieved, then. The question of greatest interest is how worth while these achievements have been and are likely to be. The motive of the Soviet Government in undertaking the programme in the first place appears to have been economic. But the usefulness of the route extends to the fields of strategy, socialisation of the native population, communist propaganda and pure science. Let us estimate, from the point of view of the Soviet Government, the success of the Northern Sea Route in each of these capacities. It is true that the evidence on which we may base conclusions is incomplete, and that the gaps in our knowledge of developments since the war are here particularly keenly felt. It must be made clear at this point that it is in fact impossible to form any estimate of the strategic advantage which may be gained by the Soviet Union, because the evidence on which to base conclusions is lacking. But there is enough information from which to draw certain conclusions on other than strategic aspects.

It cannot be said that the Northern Sea Route has been a great success economically. The national economy has not markedly benefited from its working.

Development has been speculative, in that communications have been organised before specific need for them has arisen. But it must be remembered that the economic potentialities in the shape of still untapped natural resources remain great. It is the lesser uses of the route which seem to have been the most worth while. Undoubtedly the route was an important factor in bringing the natives into the socialist system, provided good propaganda material, and furthered the cause of science. To this extent the route may fairly be said to have justified itself; but it is a very small profit on a very large outlay.

As far as we can judge, then, the route has not yet proved itself worth while in its economic role. Let us consider the prospects for the future. On the one hand it is possible that scientific and technological research will be able to bring about major improvements. During the period we have examined, Soviet impatience for results has meant that, for instance, freighter convoys were sent out at the same time as expeditions to study the hydrography of the sea lanes to be traversed; research, in fact, and practical exploitation of its results, went forward as near simultaneously as possible. Soviet scientists were themselves worried about this. Seen in this light, accidents and under-fulfilment of plans were only to be expected. Once research can get ahead and provide more accurate ice forecasts, higher-powered icebreakers, more ice-breaking freighters, information about the northern variant of the route and perhaps other variants too, then results may become really worth while. The most important outcome would be the lengthening of the shipping season. Although it may be a very long time before convoys can negotiate the ice in winter and spring, fuller utilisation of the marginal periods at the beginning and end of the season would become possible if more complete ice reports were available and the ships were ready to sail at the time required. With even a slightly longer season, the economic importance of the route would be greatly increased.

Against all this must be set another possibility. There is much evidence to support the view that the Arctic regions have been getting warmer since about 1920. This trend may continue, but on the other hand it may quite possibly be reversed. The last twenty or thirty seasons may represent a particularly favourable period for navigation. If this is the case then of course the whole perspective changes. Technical advances may be offset and far more than offset by much severer ice conditions. There is no evidence yet that the cooling off is starting; nor is enough known about past climatic fluctuations to allow any estimate of when such a trend may start. One can only say that there is a possibility that it will. And the climatic deterioration will not have to be very great to have serious consequences for the Northern Sea Route, which lies in a climatically marginal region.

In face of these possibilities the Soviet authorities may act in either of two ways. They may abandon the whole venture in view of the limited success so far achieved; or they may go ahead in the hope that the growth in scientific and technical knowledge will lead to more efficient functioning of the route. They will surely go ahead. It will not be their own propaganda which forces them to do so, for the Soviet Government is always realistic enough to drop a major project if the circumstances warrant it, however awkward the explaining away of previous propaganda may be. They will go ahead because there are reasonable grounds for belief that the route will soon be able to justify its existence economically, quite apart from any other possible justification, if its working can be further improved; and Glavsevmorput', despite mistakes and inefficiency, has learnt enough to be able to devise and use to the fullest advantage the measures necessary for such improvement.

APPENDIX I

TRAFFIC TO AND FROM THE OB' AND YENISEY, 1920–39

Year	Exports from Siberia, in metric tons unless otherwise stated	Imports to Siberia, in metric tons	Total, in metric tons	Number of freighters employed
1920	10,100	—	10,100	10
1921	13,667	8,440	22,107	5*
1922	5,837	7,790	13,627	5
1923	24	1,076	1,100	1†
1924	4,148	6,523	10,671	3
1925	5,582	7,602	13,184	4
1926	10,070	9,098	19,168	5
1927	11,114	13,314	24,428	6
1928	17,107	12,271	29,378	8
1929	60,060	13,500	73,560	26
1930	125,000	18,000	143,000	46‡
1931	49,165	14,445	63,610	16
1932	76,480§	20,283	96,763	28
1933	36,807 standards of timber‖	8,409	—	30
1934	39,931 standards	7,490	—	31
1935	49,602 standards	17,175	—	45
1936¶	139,700	6,100	145,800	not known
1937	111,700	1,800	113,500	not known
1938	174,700	3,100	177,800	45
1939	373,800	2,000	375,800	not known

Sources: For period 1920–32: ref. no. 382, supplemented by 82, 107, 155, 221, 224, 351, 365, 379; for 1933: 280; for 1934: 281; for 1935: 282; for 1936–37: 179.

  * Ref. no. 107 gives 10 ships; nos. 351 and 379 give 5.

  † Ref. no. 379 gives 3 ships. The reason for the small turnover this year was the sending of the "Curzon Note" to the Soviet Government by the British Government. In this Note satisfaction was demanded for Soviet action against British subjects, and a result of the Note was that ships which the Soviet Government was going to charter were not made available.

  ‡ Ref. no. 379 gives 40 ships, no. 107 gives 48, nos. 155 and 365 give 46.

  § Ref. no. 107 gives 83,000 tons export and 7000 tons import. The figures in the table are taken from no. 379.

  ‖ A Soviet standard of timber is 165 cu. ft., or roughly 5 cu. m., and weighs generally rather less than 3 metric tons.

  ¶ The figures for 1936–39 apply to the Yenisey only.

APPENDIX II

TRAFFIC FROM THE EAST TO THE KOLYMA, 1911–36

Year	Number of freighters employed	Imports to Siberia, in metric tons	Remarks
1911–30	Generally 1, sometimes 2, a year	Always under 1,000	—
1931	2	2,000	—
1932	6	not known	10,000 tons brought to river but much was not unloaded
1933	9	7,000	4 ships and much of the cargo from expedition of 1932
1934	4	7,000	—
1935	6	11,541	1 ship came from and returned to the west
1936	6	not known	3 ships on west to east through trip, 2 on east to west through trip

Sources: For period 1911–32: ref. no. 86; for 1933–34: 124,390; for 1935: 43; for 1936: 49.

APPENDIX III

TRAFFIC FROM THE WEST TO THE LENA, 1933–38

Year	Number of freighters employed	Imports to Siberia, in metric tons
1933	2	3,880
1934	3	7,500*
1935	5	12,828†
1936	5	13,950
1937	3‡	not known
1938	4	not known

Sources: For 1933: ref. no. 77; for 1934: 106, 227; for 1935: 239, 292; for 1936: 49, 327; for 1937–38: 373.

   * This figure is from ref. no. 227. No. 106 gives 9000 tons.

   † This figure is from ref. no. 239. No. 292 gives 14,000 tons.

   ‡ Only 3 ships arrived out of the 10 which set out. These 3 returned eastwards instead of westwards.

APPENDIX IV

FREIGHT TURNOVER ON THE NORTHERN SEA ROUTE, 1933–45

Certain figures indicating total freight turnover are available. They are often misleading because the source providing the figures seldom makes clear exactly what freight they cover: for instance, whether the total does or does not include the Kara operations, maintenance of polar stations, or shipments of coal from Spitsbergen to Murmansk or Arkhangel'sk (in some years this last would be 60% of the total). In addition, the figures generally include turnover at ports on the north part of the Bering Sea, and these are outside the Northern Sea Route as we understand it here; but this turnover is not generally large.

The figures below are thought to represent all freight movements by sea along any section of the Northern Sea Route, including Kara operations, polar station maintenance and Bering Sea traffic, but excluding coal shipments from Spitsbergen.

Year	Turnover in metric tons
1933	136,000
1934	156,000
1935	230,000
1936	271,100
1937	207,000
1938	256,000
1939	503,000
1940	491,400
1945	882,000

Sources: For 1933–36: ref. no. 393; for 1937: 73, 393; for 1938–40: 179, 237, 397; for 1945: 248. It should be pointed out that while the figures for 1933–36 are taken directly from one source, the remainder are calculated from pieces of information provided by various sources; for instance, 1945 turnover is expressed as a percentage of 1940.

APPENDIX V

SHIPPING OF LEND-LEASE GOODS FROM THE UNITED STATES TO SOVIET ARCTIC PORTS BY WAY OF THE NORTHERN SEA ROUTE, 1942–45

			Ports visited, with number of calls made at each										
Year	Number of voyages made	Total tonnage carried (gross long tons)	Arkhangel'sk	Ports on Yenisey	Mouth of Khatanga	Mouth of Anabar	Mouth of Olenek	Tiksi	Mouth of Indigirka	Mouth of Yana	Ambarchik	Pevek	Bukhta Provideniya
1942	23	64,000	—	2	3	3	1	11	1	4	9	5	1
1943	32	118,000	1	4	3	2	1	14	3	4	7	6	9
1944	34	128,000	—	4	5	2	—	15	2	5	6	6	14
1945	31	142,000	—	3	5	1	—	14	3	4	6	8	11
Totals	120	452,000	1	13	16	8	2	54	9	17	28	25	35

Source: This information is derived from the manifests of the ships concerned. Copies of these manifests are included in the files of the Foreign Economic Administration now held by the Department of State, Washington, D.C.

APPENDIX VI

TRANSLATION OF DECREE OF COUNCIL OF PEOPLE'S COMMISSARS OF 25 JANUARY 1941 OUTLINING THE STRUCTURE OF GLAVSEVMORPUT'

The Council of People's Commissars of the U.S.S.R. [Sovnarkom SSSR] decrees:

1. That it confirms the following structure of the Chief Administration of the Northern Sea Route attached to the Council of People's Commissars of the U.S.S.R. [Glavnoye Upravleniye Severnogo Morskogo Puti pri SNK SSSR]:

Political Administration [Upravleniye].
Administration of the Arctic Fleet and Ports.
Administration of the River Fleet.
Administration of Polar Aviation.
Hydrographic Administration.
Administration of Polar Stations.
Mining and Geological Administration.
Administration of Capital Construction, with a Bureau of Experts (on cost accounting basis [na khozraschete]).
Administration of Arctic Supply—"Arktiksnab" (on cost accounting basis).
Department [Otdel] of Teaching Institutions and Preparation of Cadres.
Planning and Financial Department.
Central Book-Keeping Department.
Department of Leading Cadres.
Department of Work and Pay.
Bureau of Inventions.
War Department.
Control and Inspection Group.
Secretariat of the Head, his Deputies and Collegium.
Office Management Section, with Central Archive.
Inspectorate of Security and Anti-Aircraft Defence.
Secret Cipher Department.
Group of Subsidiary Economic Enterprises.
Chief Advocate.
Chief Editor.
Arbiter.

2. That it establishes that under the direct control of the Chief Administration of the Northern Sea Route attached to the Council of People's Commissars of the U.S.S.R. are the Arctic sea shipping companies [parokhodstva], offices and ports, the river shipping organisations and offices, the polar stations and other undertakings and organisations, as mentioned in the Appendix.

3. That it permits the Head of the Chief Administration of the Northern Sea Route attached to the Council of People's Commissars of the U.S.S.R.

(a) to liquidate the Yakutsk Hydrographic Department;

(b) to reorganise the Novaya Zemlya Hydrographic Department at Arkhangel'sk and the Chukotka Hydrographic Department at Bukhta Provideniya into bases of the hydrographic service, entrusting to them piloting functions and the repair of ships of the hydrographic fleet;

(c) to reorganise the following river hydrographic detachments [otryady] into river pilot stations [lotsmeysterstva]: Khatanga, Anabar, Olenek, Lena, Indigirka and Yana;

(d) to organise an economic branch of the Arctic Research Institute with a staff of 25 in Moscow;

(e) to organise a base for laying up and carrying out repairs on the hydrographic fleet at Bukhta Tiksi, and a pilot station on the river Kolyma.

4. That it confirms for 1941 the general establishment of the central organisation of the Chief Administration of the Northern Sea Route attached to the Council of People's Commissars of the U.S.S.R. as numbering 650 persons; of this number 590 persons in the Administrations and Departments are financed by the state budget, and 60 persons in the Administrations and Departments are financed by cost accounting and special funds.

5. In connection with the present decree, that it regards as having lost effect the Statute on the Chief Administration of the Northern Sea Route attached to the People's Commissars of the U.S.S.R., confirmed by the decree of the Council of People's Commissars of the U.S.S.R. of 22 June 1936, with the following additions and amendments: §317, 1936; §184, 1938; §319, 1939; §442, 1940.

Chairman of the Council of People's Commissars of the U.S.S.R.

V. MOLOTOV

Secretary   YA. CHADAYEV

Moscow, Kremlin, 25 January 1941

APPENDIX TO DECREE OF COUNCIL OF PEOPLE'S COMMISSARS OF THE U.S.S.R. OF 25 JANUARY 1941

List of undertakings and organisations under the direct control of the Chief Administration of the Northern Sea Route attached to the Council of People's Commissars of the U.S.S.R.

Sea shipping companies [parokhodstva], offices and ports
   Arkhangel'sk Arctic sea shipping company.
   Vladivostok Arctic sea shipping company.
   Murmansk Arctic sea office.
   Port Dikson.
   Port Provideniya.
   Port Tiksi.

River shipping companies and offices
   Kolyma-Indigirka river shipping company.
   North Yakutiya river shipping company.
   Khatanga river office.

Bases and pilot stations of the hydrographic service
   Arkhangel'sk base.
   Provideniya base
   Tiksi base.
   Anabar pilot station.
   Indigirka pilot station.
   Kolyma pilot station.
   Lena pilot station.
   Olenek pilot station.
   Khatanga pilot station.
   Yana pilot station.

Polar stations

Ostrov Ayon	Mys Billingsa	Mys Uellen
Ostrov Belyy	Mys Vankarem	Mys Chaplina
Ostrov Vaygach	Mys Vkhodnoy	Mys Chelyuskina
Ostrov Vrangelya	Mys Vykhodnoy	Mys Shalaurova
Ostrov Vize	Mys Drovyanoy	Mys Shelagskiy
Ostrov Genrietty	Mys Zhelaniya	Mys Shmidta
Ostrov Domashnyy	Mys Kigilyakh	Bukhta Ambarchik
Ostrov Kotel'nyy	Mys Kolyuchino	Bukhta Tiksi
Ostrov Mostakh	Mys Leskina	Bukhta Tikhaya
Ostrova Petra	Mys Molotova	Proliv Sannikova
Ostrov Preobrazheniya	Mys Navarin	Proliv Yugorskiy Shar
Ostrov Pravdy	Mys Olovyannyy	Gorod Amderma
Ostrov Ratmanova	Mys Pestsovyy	Seleniye Gyda-Yamo
Ostrov Rudol'fa	Mys Peschanyy	Seleniye Karmakuly
Ostrov Russkiy	Mys Sterlegova	Seleniye Mare Sale
Ostrov Tyrtova	Mys Stolbovoy	Zaliv Blagopoluchiya
Ostrov Uyedineniya	Mys Serdtse Kamen'	Russkaya Gavan'
Ostrov Chetyrekhstolbovoy		

Meteorological stations

Anadyr'.

Radio stations

Bukhta Kozhevnikova.
Bukhta Provideniya.
Bukhta Ugol'naya.
Poselok Zyryanka.
Department of radio communications and hydrological and meteorological
    station at Yakutsk.

Air groups

Igarka.
Lena.
Moscow special duty.
Chukotka.

Mining undertakings and expeditions

Sangar-Khaya coal mine.
Nordvik oil expedition.
Ust'-Yeniseysk oil expedition.

Industrial undertakings

Murmansk ship repair yard.
Krasnoyarsk aircraft repair factory.
Arkhangel'sk yard for construction of wooden ships.
Kachuga yard for construction of metal ships.
Peleduy yard for construction of wooden ships.
Karacharov electro-mechanical workshop.

Construction companies [*stroitel'nyye kontory*]

Arkhangel'sk construction company.
"Sevmorput'zavodstroy" construction company at Murmansk.
"Diksonstroy" construction company.
"Providenstroy" construction company.
"Tiksistroy" construction company.

Supply and forwarding branches of "Arktiksnab"

Leningrad supply branch.
Moscow supply branch.
Sverdlovsk supply branch.
Kharkov supply agency.
Arkhangel'sk forwarding branch.
Vladivostok forwarding branch.
Irkutsk forwarding branch.
Krasnoyarsk forwarding branch.

Teaching institutions

Leningrad Hydrographic Institute.
Leningrad Hydrographic Technical School.
Murmansk Marine Technical School.
Technical School of Meteorology and Communications at Moscow.
Arctic Research Institute [Arkticheskiy Nauchno-Issledovatel'skiy In-
stitut, abbreviated to ANII] at Leningrad, with an economic branch at
Moscow.
Project research office (on cost accounting basis) in Moscow.
Publishing house.
The journal *Sovetskaya Arktika*.

Source: Ref. no. 309.

APPENDIX VII

SOVIET VESSELS FOR ICE NAVIGATION

The list below includes only vessels of more than 1000 tons displacement which are known to have been used in Arctic waters. Unless otherwise stated, the ships are thought to be still in use.

Name of ship	Where built	When completed	Displacement tonnage normal *maximum*	Gross registered tonnage	Total normal h.p.	Maximum speed in knots	Fuel	Engines	Remarks
ICEBREAKERS—ALL TYPES *Yermak*	Newcastle	1898	7,875 *10,000*	4,955	10,000	14	Coal	Triple-expansion steam reciprocating	Spent war years (1941–45) at Leningrad
Taymyr	St Petersburg	1909	1,320	—	1,200	10·5	Coal	Triple-expansion steam reciprocating	—
Vaygach	St Petersburg	1909	1,320	—	1,200	10·5	Coal	Triple-expansion steam reciprocating	Sank off mouth of Yenisey in 1918
Dobrin'ya Nikitich	Newcastle	1916	3,100 *2,460**	1,664	4,000	14	Coal	Triple-expansion steam reciprocating	Repaired in Canada during second world war
Il'ya Muromets	(?) Newcastle	(?)1916	—	—	4,000	—	—	—	Has probably not been in Arctic since 1919
Skuratov ex *Ivan Susanin*	(?) Newcastle	(?)1916	—	—	2,000	—	—	—	Has probably not been in Arctic since 1921
Krasin ex *Svyatogor*	Newcastle	1917	8,730 *10,620*	4,902	10,000	15	Coal	Triple-expansion steam reciprocating	Repaired in U.S.A. in 1942–43
Lenin ex *Aleksandr Nevskiy*	Newcastle	1917	5,074 *5,820* *6,260**	3,829	7,500	16	Coal	Triple-expansion steam reciprocating	Refitted in Mersey, 1946–47

Name	Built	Year	Tonnage			Speed	Fuel	Engine	Notes
Koz'ma Minin	—	—	—	—	6,000	—	—	—	Has probably not been in Arctic since 1919
Mikula Selyaninovich	Montreal	1916	5,000	3,165	8,000	15	Coal	Triple-expansion steam reciprocating	Taken away from White Sea by Allies when they retired in 1919
Stepan Makarov ex Knyaz' Pozharskiy	Newcastle	1916	3,570 4,600 3,150*	2,372	5,550	14·5	Coal	Triple-expansion steam reciprocating	Has probably not been in Arctic since 1919
Fedor Litke ex III Internatsional ex Canada ex Earl Grey	Barrow	1909	2,570 4,600	2,375	7,000	17	Coal	Triple-expansion steam reciprocating	Handed over to Russia during first world war. Refitted in Mersey, 1947–48
Aleksandr Sibiryakov ex Bellaventure	Glasgow	1909	2,650	1,384	2,200	13	Coal	Triple-expansion steam reciprocating	Handed over to Russia during first world war. Thought to have been lost in second world war
Georgiy Sedov ex Beothic	Glasgow	1909	3,056	1,538	2,400	13	Coal	Triple-expansion steam reciprocating	Handed over to Russia during first world war
Vladimir Rusanov ex Bonaventure	Glasgow	1909	2,600 3,217*	1,383	2,200	13·5	Coal	Triple-expansion steam reciprocating	Handed over to Russia during first world war
Malygin ex Solovey Budimirovich ex Bruce	Glasgow	1912	3,200 2,200*	1,571	2,800	12	Coal	Triple-expansion steam reciprocating	Handed over to Russia during first world war
Sadko ex Lintrose	Newcastle	1913	3,800 2,000*	1,613	2,800	12·5	Coal	Triple-expansion steam reciprocating	Handed over to Russia during first world war. Sunk in White Sea in 1916; raised and refitted in 1933
Truvor ex Sleipnir	Copenhagen	1895	1,450 2,000	999	1,151	13	Coal	Triple-expansion steam reciprocating	Thought to have been lost in second world war

Name of ship	Where built	When completed	Displacement tonnage normal / *maximum*	Gross registered tonnage	Total normal h.p.	Maximum speed in knots	Fuel	Engines	Remarks
Davydov ex Krasnyy Oktyabr' ex Nadezhnyy	Copenhagen	1898	1,700 / *2,170*	1,212	2,475	14·4	Coal	Double-expansion	—
Iosif Stalin	Leningrad	1988	9,300 / *11,000*	4,866	10,000	14·5	Coal	Triple-expansion steam reciprocating	Repaired in U.S.A. during second world war
Lazar' Kaganovich	Nikolayev	1989	9,300 / *11,000*	4,866	10,000	14·5	Coal	Triple-expansion steam reciprocating	—
Vyacheslav Molotov	Leningrad	(?)1941	9,300 / *11,000*	4,866	10,000	14·5	Coal	Triple-expansion steam reciprocating	Spent war years (1941–45) at Leningrad
Anastas Mikoyan	Nikolayev	(?)1941	9,300 / *11,000*	4,866	10,000	14·5	Coal	Triple-expansion steam reciprocating	Repaired in U.S.A. during second world war
Kapitan Belousov ex Severnyy Veter(?)	San Pedro, California	(?)1944	5,300 / *6,515*	—	10,000	16·5	Oil	Diesel-electric	Transferred to U.S.S.R. under lend-lease
Severnyy Polyus	San Pedro, California	1944	5,300 / *6,515*	—	10,000	16·5	Oil	Diesel-electric	Transferred to U.S.S.R. under lend-lease
Admiral Makarov	San Pedro, California	(?)1944	5,300 / *6,515*	—	10,000	16·5	Oil	Diesel-electric	Transferred to U.S.S.R. under lend-lease. One of the three ships was returned to U.S.A. in December 1949, the other two in December 1951
Montcalm	Glasgow	1904	3,270	1,482	3,225	14	Coal	Triple-expansion steam reciprocating	Transferred to U.S.S.R. from Canada in 1942
Projected icebreaker	—	—	7,700	—	12,000	16·2	Oil	Diesel-electric	—

Projected icebreaker	—	—	10,650	—	18,000	—	Oil	Diesel-electric	—
Projected icebreaker	—	—	15,700	—	24,000	19	Oil	Diesel-electric	—
Projected icebreaker	—	—	16,750	—	24,000	19	Oil	Turbo-electric	—
FREIGHTERS SPECIALLY DESIGNED FOR ICE									
Chelyuskin	Copenhagen	1933	6,400	—	2,400	11·5	Coal	Triple-expansion steam reciprocating	Sank north of Bering Strait in 1934
Dezhnev	Leningrad	1938	6,530	2,140†	2,500	11·5	Coal	Triple-expansion steam reciprocating	—
Levanevskiy	Leningrad	1940	6,530	2,140†	2,500	11·5	Coal	Triple-expansion steam reciprocating	—
Nenets (tanker)	Yokohama	1937	—	1,681	—	—	—	—	—
Yukagir (tanker)	Yokohama	1937	—	1,681	—	—	—	—	—
Seamorput' II (projected freighter)	—	—	8,700	3,250†	3,000	12·5	Coal	Triple-expansion steam reciprocating	—
Seamorput' II (projected freighter)	—	—	8,700	3,880†	3,000	12·5	Oil	Diesel-electric	—
Projected tanker	—	—	7,920	3,850†	4,400	14	Oil	Motor ship	—

Sources: Ref. nos. 94, 158, 173, 174, 240, 360. I. V. Vinogradov (360) and H. F. Johnson (94) have been considered the most reliable sources. Their figures generally agree, and have been taken as the basis for the table above. Where two or more sources overlap, there are frequent discrepancies. The differences do not generally exceed 250 tons displacement or 500 h.p., and it has not been thought worth while to indicate these. The displacement tonnages given in Jane's Fighting Ships (174), however, sometimes differ considerably from those given by Vinogradov and Johnson, and in these cases both figures have been given, Jane's being marked with an asterisk. Gross registered tonnage is taken from English sources, as Soviet tables do not normally give it. Soviet sources do mention load capacity [chistaya gruzopod''yemnost'] in tons; this, which is marked with a dagger in the table above, is probably equivalent to deadweight tonnage.

APPENDIX VIII

PROXIMATE ANALYSES OF COAL FOUND IN THE SOVIET ARCTIC

Location	Moisture percentage	Ash percentage	Sulphur percentage	Volatiles percentage	Calories
Spitsbergen					
Barentsburg	1·21– 3·41	7·45–13·85	0·87–3·63	33·44–38·4	6260–8416
Pyramiden	1·18– 1·31	11–22	0·5 –1·5	28–59	6650–7800
Pechora Basin					
Vorkuta	1·63– 3·9	3·68–13·5	0·49–0·88	22·05–31·5	7226–7637
Zemlya Frantsa-Iosifa	17·7–47·9	1·66– 6·24	—	—	3318–5369
Yugorskiy Poluostrov					
Liur	1·3 – 5	24·2 –30·7	0·16–1·16	4·61– 8·9	5251–6478
Korotaikha	0·93– 1·19	31·6 –36·38	0·49–0·5	15·31–15·4	5187–5607
Tungus Basin					
South part (lat. 57°–62° N.)	4·86–20·05	4·65–24·9	0·87–0·99	17·1 –57·4	—
Central part (lat. 65°–68° N.)	0·62– 8·1	5·15–14·7	0·28–0·79	3·36–28·8	—
North part (lat. 68°–71° N.)					
Noril'sk	2·2 – 3·72	9·4 –10·9	0·68–0·74	17·68–22·2	7440–7522
Kotuy	4·32–18·74	4·3 –12·87	0·25–1·22	32·5 –44·7	5958–6896
Taymyr Basin					
Reka Lemberova	1·08– 5·27	10·14–23·94	0·1 –3·8	3·73– 5·42	6264–6862
Reka Uboynaya	2·33– 2·39	5·44– 5·69	0·28–0·64	20·24–22·91	7495–7529
Reka Pyasina, right bank	2·04– 4·54	6·07– 9·04	0·49–0·58	4·82– 9·54	7800–8250
Reka Pyasina, left bank	1·6 – 2·4	5·6 –13·9	0·5 –0·64	18·7 –20·9	8290–8480
Reka Tarey	1·74– 1·96	6·74–10·87	0·31–0·38	32·7 –33·1	6967–7898
Bukhta Kozhevnikova	10·35–15·77	4·89– 9·78	0·25–0·63	35·91–39·45	5185–6220

Nordvik (Yurung-Tumus)*	7–22	5–15	0·28–0·72	—	4360–5304
Lena Basin					
Olenek humus coal	11·5	3·6	0·53	34·8	—
Olenek boghead coal	0·55	2·3	0·49	90·8	—
Bulun	3·13	24·16	0·55	36–37	5614
Sangar-Khaya	4·27	10·04	0·38	52·65	6118
Kangalasskoye	19·5	6·5	0·3	51·55	5070
Zhigansk	11·03	12·65	0·4	46·6	5850
Tiksi†	15·6	3·66	1·01	46·86	—
Kolyma-Indigirka Basin					
Reka Selegnyakh	5·19	17·01	0·7	48·65	—
Zyryanka‡	3·03	8·58	0·29	36	6915
Anadyr'					
Telegraficheskoye	13·16–19	4·7 – 9·5	0·12–1·15	44–61	4370–5400
Reka Ugol'naya	17·28	3·6	0·3	57·6	—
Bukhta Ugol'naya	1·17–4·45	5–25	—	35–45·8	7508–8178

Sources: Principally ref. no. 183, which is based on information available up to early 1939. Cases where other sources have been used, or where other sources conflict with the main source, are indicated below. Fixed carbon percentage is not normally given in Soviet proximate analyses. For map, see p. 80.

* These figures taken from ref. no. 110.

† Ref. no. 41 gives 8·2–19·5 % moisture, 3·9–9·57 % ash, 0·52–2·56 % sulphur, 42·7–48·4 % volatiles, 3196–5053 calories.

‡ Ref. no. 204 gives 2·96 % moisture, 8·44 % ash, 0·48 % sulphur, 28·9 % volatiles, 8305 calories.

APPENDIX IX

COASTAL POLAR STATIONS IN THE SOVIET ARCTIC,
1896–1948

Stations on the south shore of the Barents Sea, i.e. on the mainland of European Russia, are not included in this list since their functions are not closely related to the Northern Sea Route.

Not all stations have necessarily been working continuously since the date of their opening. Some were closed when relief could not reach them and were reopened later.

Stations which functioned only for a short period, e.g. those set up for the International Polar Years of 1882–83 and 1932–33, or by expeditions, are not included in the list.

For the location of these stations, see Map 8, p. 137.

Name of station	Date regular observations commenced	Remarks
Barents Sea		
1. Barentsburg	1932*	—
2. Bukhta Tikhaya	1929*	—
3. Ostrov Rudol'fa	1932*	—
4. Bugrino	1925*	—
5. Severnaya	— *	—
6. Russkaya Gavan'	1932*	—
7. Malyye Karmakuly	1896*	—
8. Mys Stolbovoy	1934*	—
9. Ozernaya	1935	Probably closed before the war
10. Guba Belush'ya	— *	—
11. Mys Greben'	1934	Mys Greben' and Bukhta Varneka
12. Bukhta Varneka	1930	may be identical, although one authority (96) distinguishes between them
13. Khabarovo	— *	—
Kara Sea		
14. Mys Zhelaniya	1931*	—
15. Zaliv Blagopoluchiya	1936*	—
16. Matochkin Shar	1923*	—
17. Mys Vykhodnoy	1934*	—
18. Zaliv Abrosimova	— *	—
19. Ostrov Vaygach	1914*	—
20. Yugorskiy Shar	1913*	—
21. Amderma	1934*	—
22. Karskaya Guba	1933*	—
23. Mare Sale	1914*	—
24. Ostrov Belyy	1933*	—
25. Mys Drovyanoy	1932*	—
26. Tambey	— *	—
27. Se-Yakha	— *	—
28. Novyy Port	1924*	—
29. Gyda-Yamo	1932*	—
30. Mys Leskina	1934*	—

Name of station	Date regular observations commenced	Remarks
Kara Sea (cont.)		
31. Ostrov Diksona	1915*	—
32. Gol'chikha	?1934	The station at Gol'chikha may have been moved to Sopochnaya Korga
33. Sopochnaya Korga	— *	—
34. Ust'-Port	1920*	—
35. Mys Vkhodnoy	— *†	—
36. Mys Sterlegova	1934*	—
37. Ust'ye Taymyry	1935*	—
38. Ostrov Tyrtova	1940*	At first manned during summer only
39. Ostrov Russkiy	1935*	—
40. Ostrov Pravdy	1940‡	—
41. Ostrova Geyberga	1940*	At first manned during summer only
42. Mys Chelyuskina	1932*	—
43. Mys Neupokoyeva	— †	—
44. Mys Olovyannyy	1935*	—
45. Ostrov Domashniy	1930*	—
46. Ostrov Uyedineniya	1934*	—
47. Ostrov Vize	— †‡	—
Laptev Sea		
48. Mys Peschanyy	— ‡	—
49. Ostrov Malyy Taymyr	— *	—
50. Ostrova Samuila, renamed in 1935 Ostrova Komsomol'skoy Pravdy	1933–39	—
51. Ostrov Andreya	1940*	—
52. Ostrova Petra	1940*	—
53. Bukhta Pronchishchevoy	1935–37	—
54. Ostrov Preobrazheniya, renamed Ostrov Vstrechnyy	1934*	—
55. Nordvik	1932–38	—
56. Mys Kosistyy	1937*	—
57. Ust'-Olenekskoye	1940*	—
58. Mys Barkin Stan	— *	—
59. Bukhta Tiksi	1932*	Later also called Sogo
60. Ostrov Mostakh	1935*	—
61. Ostrov Kigilyakh	1934*	—
62. Sannikovo	1940*	—
63. Ostrov Kotel'nyy	1933*	—
East Siberian Sea		
64. Mys Shalaurova	1928*	—
65. Mys Pestsovyy	— ‡	—
66. Ostrov Genrietty	1937*	—
67. Ostrov Chetyrekhstolbovoy	1933	—
68. Mys Medvezhiy	1935–39	Also called Ust'ye Kolymy. Moved to Ambarchik in 1939
69. Ambarchik	1939*	—
70. Ostrov Rauchua	— *‡	—
71. Ostrov Ayon	1939*	At first manned during summer only
72. Pevek	— *	—
73. Mys Shelagskiy	1934*	—
74. Mys Billingsa	1935*	—

Name of station	Date regular observations commenced	Remarks
Chukchi Sea and Bering Strait		
75. Mys Shmidta	1932*	—
76. Ostrov Vrangelya	1926*	—
77. Mys Litke	1940	Possibly only a temporary station
78. Guba Kolyuchinskaya	1934*	Probably identical with station called Dzhinretlen
79. Ostrov Kolyuchin	— *	Probably identical with station called Kolyuchino
80. Mys Vankarem	1934*	—
81. Mys Serdtse-Kamen'	1933*	—
82. Uelen	1928*	—
83. Mys Dezhneva	1932	Also worked 1916–18. May have closed in 1937
84. Zaliv Lavrentiya	?1934‡	—
85. Mys Chaplina	— *	—
86. Ostrov Ratmanova	?1940*	—
87. Bukhta Provideniya	?1934*	—

Sources: Ref. nos. 67, 100, 157, 159, 243, 252, 309.

   * Known to be functioning in 1948 (ref. no. 157).

   † Lend-lease goods were addressed to these stations in 1943 or 1944 (information from ships' manifests; see Appendix V).

   ‡ Station under Glavsevmorput' control in 1941 (ref. no. 309). This information is included only when the fact that a station was functioning in 1941 is not apparent from any other source.

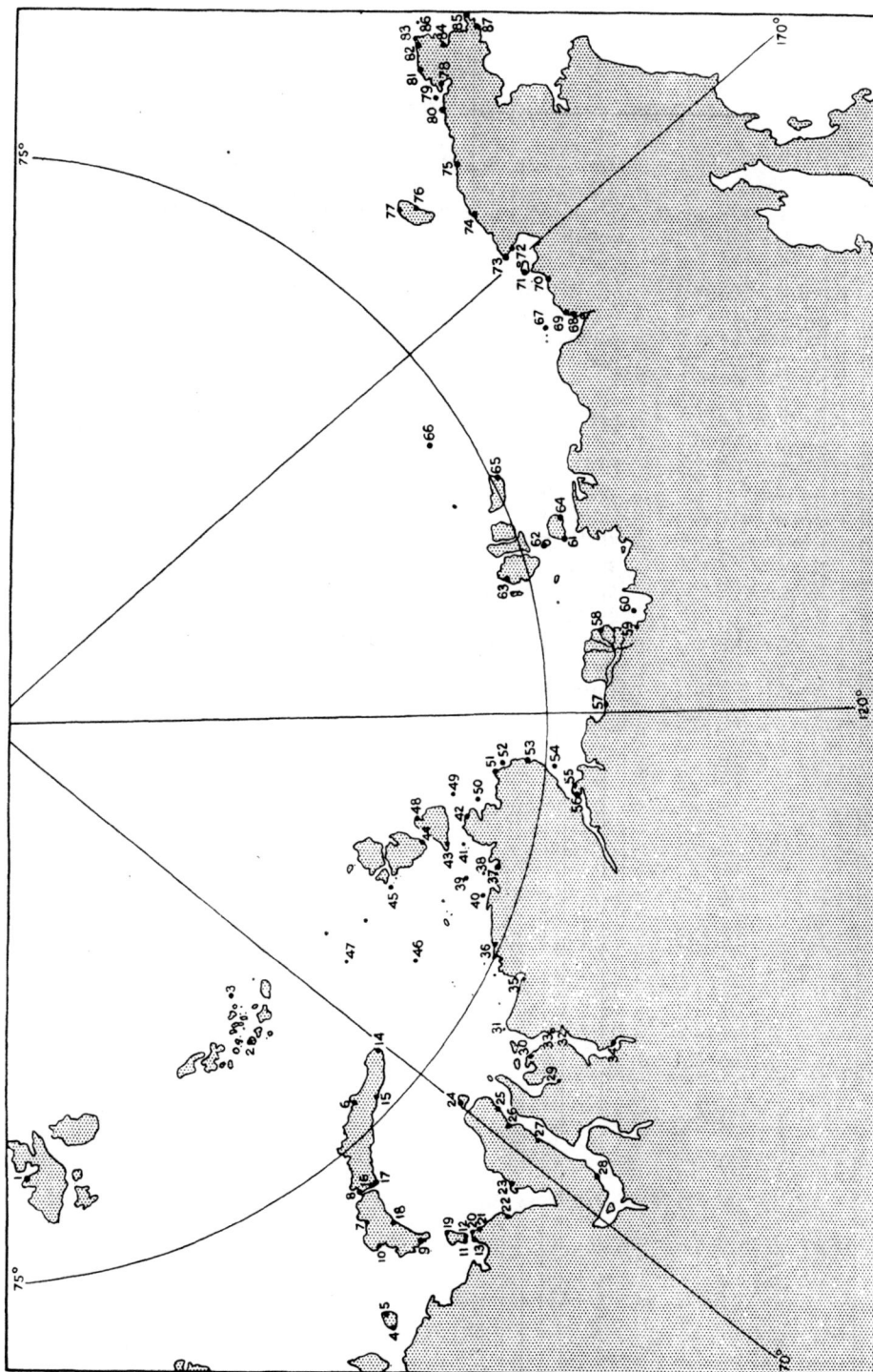

Map 8. Coastal polar stations (the key to the numbers is on the preceding pages)

GLOSSARY OF RUSSIAN TERMS USED IN THE TEXT

A.S.S.R.	Autonomous Soviet Socialist Republic.
Bukhta	bay.
Glavsevmorput'	Chief Administration of the Northern Sea Route.
Gorod	town.
Guba	bay.
Komseverput'	Committee of the Northern Sea Route.
Komsomol	Young Communist League.
Khrebet	mountain range.
Kray	province.
More	sea.
Mys	cape.
Narkomvod	People's Commissariat for Water Transport.
Oblast'	province.
Ostrov(a)	island(s).
Poluostrov	peninsula.
Poselok	settlement.
Proliv	strait.
Reka	river.
Seleniye	settlement.
Sovnarkom	Council of People's Commissars.
Zaliv	gulf.
Zemlya	land.

REFERENCES

ABBREVIATIONS

Bol.Sov.Ent. = *Bol'shaya Sovetskaya Entsiklopediya* [*Large Soviet Encyclopaedia*], Moscow, 65 Vols., 1926–46.

Byull.Ark.Inst. = *Byulleten' Arkticheskogo Instituta* [*Bulletin of the Arctic Institute*], Leningrad, 1931–36.

Geog.Journal = *Geographical Journal*, London, 1893–. Published by the Royal Geographical Society.

Izv.Arkh.Ob.iz.Russ.Sev. = *Izvestiya Arkhangel'skogo Obshchestva izucheniya Russkogo Severa* [*News of the Arkhangel'sk Society for the study of the Russian North*], Arkhangel'sk, 1909–?19.

Izv.Vse.Geog.Ob. = *Izvestiya Vsesoyuznogo Geograficheskogo Obshchestva* [*News of the All-Union Geographical Society*], Leningrad, 1940–.

Mor.Flot = *Morskoy Flot* [*Merchant Fleet*], Moscow, 1941–. Published monthly by Ministerstvo Morskogo Flota SSSR [Ministry of the Merchant Fleet of the U.S.S.R.].

Nedra Ark. = *Nedra Arktiki* [*Mineral Resources of the Arctic*], Moscow, Leningrad, 1946–. Published by Gorno-Geologicheskoye Upravleniye Glavsevmorputi [Mining and Geological Administration of Glavsevmorput'].

Prob.Ark. = *Problemy Arktiki* [*Problems of the Arctic*], Leningrad, 1937–. Published by Arkticheskiy Institut [Arctic Institute].

Sev.Aziya = *Severnaya Aziya, Obshchestvenno-Nauchnyy Zhurnal* [*North Asia, a Journal of Social Science*], Moscow, 1925–30. Published by Obshchestvo Izucheniya Urala, Sibiri i Dal'nego Vostoka [Society for the Study of Ural, Siberia and the Far East]; Komitet Sodeystviya Narodnostyam Severnykh Okrain [Committee for Assisting the Nationalities of the Northern Territories]; Glavnauka [Chief Administration of Scientific Institutions].

Sev.Mor.Put' = *Severnyy Morskoy Put'. Sbornik Statey po Gidrografii i Moreplavaniyu* [*Northern Sea Route. Collected papers on Hydrography and Navigation*], Leningrad, Moscow, 1934–. Published by Gidrograficheskoye Upravleniye Glavsevmorputi [Hydrographic Administration of Glavsevmorput'].

Sib.Sov.Ent. = *Sibirskaya Sovetskaya Entsiklopediya* [*Siberian Soviet Encyclopaedia*], Novosibirsk, 3 vols., 1929–32. Only 3 vols. issued out of intended (?)6.

Sov.Ark. = *Sovetskaya Arktika* [*Soviet Arctic*], Moscow, 1935–. Published by Glavsevmorput'.

Sov.Aziya = *Sovetskaya Aziya, Obshchestvenno-Nauchnyy Zhurnal* [*Soviet Asia, a Journal of Social Science*], ?Moscow, 1930–31. Continuation of *Severnaya Aziya*.

Sov.Sev. = *Sovetskiy Sever* [*Soviet North*], Moscow, 1929–35. Published by Komitet Severa [Committee of the North].

Tekh.Mol. = *Tekhnika-Molodezhi* [*Technical Knowledge for Youth*], Moscow, 1933–. Published by Tsk VLKSM [Central Committee of Komsomol].

Trudy Ark.Inst. = *Trudy Arkticheskogo Instituta* [*Transactions of the Arctic Institute*], Leningrad, 1930–.

Trudy N-I.Inst.Pol.Zem.Seriya Prom.Khoz. = *Trudy Nauchno-Issledovatel'skogo Instituta Polyarnogo Zemledeliya, Zhivotnovodstva i Promyslovogo Khozyaystva. Seriya Promyslovoye Khozyaystvo* [*Transactions of the Research Institute of Polar Agriculture, Stock Raising and Hunting Economics. Hunting Economics Series*], Leningrad, (?)1938–.

Za Ind.Sov.Vostoka = *Za Industrializatsiyu Sovetskogo Vostoka* [*For the Industrialisation of the Soviet East*], Moscow, 1932–(?). Continuation of *Sovetskaya Aziya*.

(1) AARSGAARD, G. "Svalbard etter krigen". *Polarboken*, Oslo, 1949, p. 27–52. (Reference to p. 51.)

(2) ALEKSEYEV, N. N. "Gidrologicheskiye raboty v arktike v III pyatiletii" [Hydrological work in the Arctic during the third five-year plan], *Prob.Ark.* No. 10/11, 1939, p. 102–04.

(3) ALIMOV, I. V. "Dikson i Tiksi" [Dikson and Tiksi], *Sov.Ark.* No. 3, 1939, p. 42–48.

(4) ALIMOV, I. V. "Ispravit' oshibki proshlogodney navigatsii" [Correct the mistakes of last year's navigation season], *Sov.Ark.* No. 1, 1938, p. 18–32.

(5) ALIMOV, I. V. "Nash rechnoy flot v 1936 godu" [Our river fleet in 1936], *Sov.Ark.* No. 5, 1936, p. 35–43.

(6) ALIMOV, I. V. "Port Dikson nakanune navigatsii" [Port Dikson on the eve of the navigation season], *Sov. Ark.* No. 7, 1939, p. 39–46.

(7) ALIMOV, I. V., and BEREZIN, V. V. "Rechnoy transport Glavsevmorputi v navigatsiyu 1937 goda" [River transport of Glavsevmorput' in the navigation season of 1937], *Sov.Ark.* No. 4, 1937, p. 76–85.

(8) "Arkticheskiy port Dikson" [The arctic port of Dikson], *Sov.Ark.* No. 11, 1940, p. 90.

(9) ARMSTRONG, T. E. "The voyage of the *Komet* along the Northern Sea Route, 1940". *Polar Record*, Cambridge, Vol. 5, No. 37/38, 1949, p. 291–95.

(10) ARNGOL'D, E. *Po zavetnomu puti [Along the promised road]*. Moscow, Leningrad, Gosizdat, 1929. 196 p.

(11) ASHBY, E. *Scientist in Russia*. London, Penguin Books, 1947. 252 p. (Reference to p. 16–19.)

(12) *Ibid.* (p. 118).

(13) BADIGIN, K. S. *Na korable "Georgiy Sedov" cherez ledovityy okean. Zapiski kapitana [In the ship "Georgiy Sedov" across the Arctic Ocean. The captain's notes]*. Moscow, Leningrad, Izdatel'stvo Glavsevmorputi, 1941. 606 p. (Contains some preliminary scientific results.)

(14) BALZAK, S. S., VASYUTIN, V. F., and FEIGIN, YA. G., ed. *Economic geography of the U.S.S.R.* New York, Macmillan, 1949. xiv + 620 p. (Russian edition published 1940.) (Reference to p. 266.)

(15) BARABANOV, N. "Rabota ledokolov s nosovymi vintami" [Performance of icebreakers with forward screws], *Mor.Flot*, No. 7, 1948, p. 24–27.

(16) BASHMAKOV, P. I. "Pervaya russkaya morskaya ekspeditsiya k ust'yu reki Obi (1734–39 godov) i dal'neysheye razvitiye moreplavaniya v ust'ye sibirskikh rek" [The first Russian sea-going expedition to the mouth of the river Ob' (1734–39) and subsequent development of sea-going navigation to the mouths of Siberian rivers], *Sev.Mor.Put'*, No. 13, 1939, p. 43–64.

(17) BELOUSOV, M. P. "Tactical principles of navigation in ice." Reverse side of *U.S. Hydrographic Office Chart No. 2600*, 1947. (Translation of paper probably written in about 1940.)

(18) BERDNIKOV, N. "O khozraschete na sudakh" [On cost accounting in ships], *Sov.Ark.* No. 9, 1939, p. 22–24.

(19) BEREZIN, V. V. "Rechnaya navigatsiya 1937 goda" [The river navigation season of 1937], *Sov.Ark.* No. 3, 1938, p. 17–24.

(20) BERG, L. S. *Otkrytiye Kamchatki i ekspeditsii Beringa, 1725–1742. Tret'ye izdaniye [The discovery of Kamchatka and Bering's expeditions, 1725–1742. 3rd edition]*. Moscow, Leningrad, Izdatel'stvo Akedemii Nauk SSSR, 1946. 379 p. (Reference to p. 27–38.)

(21) *Ibid.* (p. 297).

(22) BERGAVINOV, S. A. "Stakhanovskoye dvizheniye—v arktiku" [Introduce the Stakhanov movement to the Arctic], *Sov.Ark.* No. 1, 1936, p. 15–19.

(23) BETEKHTIN, A. G., and others. *Kurs mestorozhdeniy poleznykh iskopayemykh. 2-e izdaniye, pererabotannoye i dopolnennoye [Primer of useful mineral deposits. 2nd edition, revised and augmented]*. Moscow?, Gostoptekhizdat, 1946. 592 p. (Reference to p. 198.)

(24) BOL'SHAKOV, V. S. "Gibel' ekspeditsionnogo sudna *Akademik Shokal'skiy*" [Loss of the expedition ship *Akademik Shokal'skiy*], *Prob.Ark.* No. 1, 1944, p. 157–59.

(25) BUDTOLAYEV, N. M. "Iz Igarki do Diksona" [From Igarka to Dikson], *Sov.Ark.* No. 12, 1940, p. 12–19.

(26) BUDTOLAYEV, N. M. "Stroitel'stvo Igarskogo porta" [Construction of Igarka port], *Sov.Ark.* No. 3, 1935, p. 21–23.

(27) BUTLER, G. G., and FLETCHER-VANE, F. P. *The sea route to Siberia*. London, Anglo-Siberian Trading Syndicate, 1890. 31 p. (Reference to p. 8.)

(28) BUYNITSKIY, V. KH., ed. *Trudy dreyfuyushchey ekspeditsii Glavsevmorputi na ledokol'-nom parokhode "G. Sedov"*, 1937–40 [*Transactions of the drifting expedition of Glavsevmorput' in the icebreaking ship "G. Sedov"*, 1937–40. Moscow, Leningrad, Izdatel'stvo Glavsevmorputi, 1946. Tom 3. (Only this volume out of the intended four has so far (1950) appeared in western Europe.)

(29) CHAPLYGIN, YE. I. "Gidrologicheskiye nablyudeniya polyarnykh stantsiy" [Hydrological observations of polar stations], *Prob.Ark.* No. 1, 1940, p. 94–96.

(30) CHERNENKO, M. B., and SELYAVINA, T. D., ed. *Polyarniki v otechestvennoy voyne* [*Polar workers in the fatherland war*]. Moscow, Leningrad, Izdatel'stvo Glavsevmorputi, 1945. 260 p. (Reference to p. 7 of introduction by I. D. Papanin.)

(31) *Ibid.* (p. 7–8 of introduction by I. D. Papanin).

(32) CHIZHIKOV, V. "Gidroledorez" [Hydraulic ice cutter], *Tekh.Mol.* No. 4, 1946.

(33) DADYKIN, V. "Indigirskaya ekspeditsiya" [Indigirka expedition], *Sov.Ark.* No. 7, 1937, p. 45–49.

(34) DALLIN, D. J., and NICOLAEVSKY, B. I. *Forced labour in Soviet Russia*. London, Hollis and Carter, 1948. xv + 331 p. (Reference to p. 61–72, 108–46.)

(35) DAVYDOV, L. K. "Vskrytiye rek arkticheskoy i subarkticheskoy zony SSSR" [Break-up of the ice on the rivers of the arctic and sub-arctic zone of the U.S.S.R.], *Prob.Ark.* No. 1, 1939, p. 15–31. (Reference to p. 24–31.)

(36) DE LONG, EMMA, ed. *The voyage of the Jeanette*. Boston, Houghton Mifflin, 1884, 2 vols.

(37) DEMIDOV, S. "Arkticheskomu flotu—mestnoye toplivo" [Local fuel for the Arctic fleet], *Sov.Ark.* No. 6, 1939, p. 25–31.

(38) "Deyatel'nost' Gos. Akts. Ob-va Komseverput' v 1932 godu" [Activity of the State Stock Company Komseverput' in 1932], *Byull.Ark.Inst.* No. 2, 1932, p. 27–28.

(39) DMITRIYEV, V. I. "Ryby i rybnyy promysel v nizov'yakh reki Yeniseya" [Fish and the fishing industry on the lower reaches of the river Yenisey], *Trudy N-I.Inst.Pol. Zem.Seriya Prom.Khoz.* No. 16, 1941, p. 7–36. (Reference to p. 7–8.)

(40) *Ibid.* (p. 34–35.)

(41) DOLGOPOLOV, N. N. "Khimicheskaya kharakteristika i puti ispol'zovaniya Tiksinskogo uglya" [Chemical characteristics and ways of using coal from Tiksi], *Nedra Ark.* No. 2, 1947, p. 113–17.

(42) DROGAYTSEV, D. A. "Desyat' let sluzhby pogody v arktike" [Ten years of the weather service in the Arctic], *Prob.Ark.* No. 2, 1944, p. 115–19.

(43) DUPLITSKIY, D. S. "Pokhod *Krasina*" [The voyage of the *Krasin*], *Sov.Ark.* No. 2, 1936, p. 36–49.

(44) DZERDZEYEVSKIY, B. L. "Organizatsiya raboty sluzhby pogody v arktike" [Organisation of the work of the weather service in the Arctic], *Byull.Ark.Inst.* No. 10, 1934, p. 368–70.

(45) "Ekspeditsiya a/o 'Komseverput'' dlya issledovaniy reki Pyasiny" [The expedition of the stock company "Komseverput'" to study the river Pyasina], *Byull.Ark.Inst.* No. 8/10, 1932, p. 201.

(46) "Ekspeditsiya Instituta po izucheniyu Severa na Zemlyu Frantsa-Iosifa v 1930 godu" [The expedition of the Institute for the Study of the North (Institut po izucheniyu Severa) to Zemlya Frantsa-Iosifa (Franz Josef Land) in 1930], *Byull.Ark.Inst.* No. 1/2, 1931, p. 4–6.

(47) EYNOR, O. L. "Antratsitovyye ugli Karskogo poberezh'ya Yugorskogo Poluostrova" [Anthracite coals of the Kara Sea shore of Yugorskiy Poluostrov], *Prob.Ark.* No. 7/8, 1939, p. 53–69. (Reference to p. 53–54.)

(48) EYSSEN, R. "Komet umfährt Sibirien", *Hamburger Fremdenblatt*, Hamburg, 3 and 4 April 1943.

(49) GAKKEL', YA. YA. "Arkticheskaya navigatsiya 1936 goda" [Arctic navigation season of 1936], *Byull.Ark.Inst.* No. 10/11, 1936, p. 445–50.

(50) GAKKEL', YA. YA. "Arkticheskaya navigatsiya 1937 goda" [Arctic navigation season of 1937], *Prob.Ark.* No. 1, 1938, p. 117–34.

(51) GAKKEL', YA. YA. "Arkticheskaya navigatsiya 1938 goda" [Arctic navigation season of 1938], *Prob.Ark.* No. 5/6, 1938, p. 135–42; No. 1, 1939, p. 72–77.

(52) *Ibid.* (p. 135, 72).

(53) GAKKEL', YA. YA. "Lotsiya Vostochnosibirskogo morya 1939" [Pilot of the East Siberian Sea, 1939], *Prob.Ark.* No. 9, 1939, p. 103–06. (Review of book named in title.)

(54) GAKKEL', YA. YA. "Oshibki v arkticheskoy navigatsii 1937 goda" [Mistakes in the Arctic navigation season of 1937], *Sov.Ark.* No. 3, 1938, p. 28–40.

(55) *Ibid.* (p. 30).

(56) GAKKEL', Ya. Ya. *Za chetvert' veka* [*For a quarter of a century*]. Moscow, Leningrad, Izdatel'stvo Glavsevmorputi, 1945. 109 p. (Reference to p. 26.)

(57) *Ibid.* (p. 40).

(58) *Ibid.* (p. 63).

(59) *Ibid.* (p. 70).

(60) *Ibid.* (p. 75).

(61) *Ibid.* (p. 80).

(62) *Ibid.* (p. 88).

(63) *Ibid.* (p. 89).

(64) GEDROYTS, N. A. "Perspektivy neftenosnosti severa Sibiri" [Oil-bearing prospects in the north of Siberia], *Nedra Ark.* No. 1, 1946, p. 9–14.

(65) GEDROYTS, N. A. "Ust'-Yeniseyskiy port i perspektivy yego neftenosnosti" [Ust'-Yeniseyskiy Port and its oil-bearing prospects], *Prob.Ark.* No. 3, 1940, p. 110–23.

(66) GEORGIYEVSKIY, N. "Kratkiye svedeniya o nauchnoy rabote na seti morskikh polyarnykh stantsiy" [Short notes on the scientific work of the network of coastal polar stations], *Prob.Ark.* No. 6, 1939, p. 70.

(67) GEORGIYEVSKIY, N. "Navigatsionnyye gidrometeorologicheskiye punkty v 1940 g." [Hydrological and meteorological bases used during the navigation season in 1940], *Sov.Ark.* No. 7, 1940, p. 68–72.

(68) GEORGIYEVSKIY, N. P. "Radiostantsii Karskogo morya" [Radio stations of the Kara Sea], *Izv.Arkh.Ob.iz.Russ.Sev.* No. 5, 1916, p. 198–204.

(69) "Gidroledorez" [Hydraulic ice cutter], *Prob.Ark.* No. 3, 1939, p. 81.

(70) GOLDER, F. A. *Russian expansion on the Pacific.* Cleveland, Arthur H. Clark, 1914. 368 p. (Reference to p. 67–95.)

(71) GOLDMAN, BOSWORTH. *Red road through Asia.* London, Methuen, 1934. xii+277 p.

(72) GOMOYUNOV, K. A. "Gidrologicheskiye issledovaniya v sovetskoy arktike za 25 let (1920–45)" [Hydrological studies in the Soviet Arctic during 25 years (1920–45)], *Izv.Vse.Geog.Ob.* Tom 77, No. 6, 1945, p. 328–40. (Contains selected list of works published on this subject. English summary of this paper appears in *Polar Record*, Vol. 5, No. 37/38, 1949, p. 355–60.)

(73) GONTSOV, D. YA. "Nekotoryye itogi 1937 goda" [Some results of 1937], *Sov.Ark.* No. 4, 1938, p. 12–18.

(74) GOTSKIY, M. B. "V vostochnom sektore arktiki" [In the eastern sector of the Arctic], in ZUBOV *et al.*, No. 405 below, p. 164–74.

(75) GRDZELOV, L. "Zadachi GGU v chetvertoy stalinskoy pyatiletke" [Tasks of the Mining and Geological Administration in the fourth Stalin five-year plan], *Nedra Ark.* No. 1, 1946, p. 6–8.

(76) GRUBER, RUTH. *I went to the Soviet Arctic.* London, Gollancz, 1939. 380 p.

(77) GURARI, G. N. "GUSMP" [Glavsevmorput'], *Sov. Sev.* No. 6, 1934, p. 25–33.

(78) *A handbook of Siberia and Arctic Russia.* London, Admiralty—Intelligence Department, 1918. 3 vols. (Reference to vol. 2, p. 87.)

(79) HOEL, A. "Coal-mining in Svalbard", *Polar Record*, Cambridge, Vol. 2, No. 16, 1938, p. 74–85.

(80) "Indigirskaya ekspeditsiya Narkomvoda" [Indigirka expedition of the People's Commissariat for Water Transport], *Byull.Ark.Inst.* No. 8, 1931, p. 153–54.

(81) IOFFE, S. *The Northern Sea Route as a transport problem.* Moscow, 1936. 27 p. (Reference to p. 3.)

(82) *Ibid.* (p. 10).

(83) *Ibid.* (p. 24).

(84) *Ibid.* (p. 24–25).

(85) ITIN, V. *Morskiye puti sovetskoy arktiki* [*Sea routes of the Soviet Arctic*]. Moscow, Izdatel'stvo Sovetskaya Aziya, 1933. 111 p.

(86) *Ibid.* (p. 87–88).

(87) "Izbranniki naroda—polyarniki" [Polar workers who have been elected by the people], *Sov.Ark.* No. 12, 1937, p. 22–25.

(88) *Izvestiya*, Moscow, 9 April 1946.

(89) JOHNSON, HENRY. *The life and voyages of Joseph Wiggins F.R.G.S.* London, John Murray, 1907. xxiv + 396 p.

(90) *Ibid.* (p. 197).

(91) *Ibid.* (p. 300).

(92) *Ibid.* (p. 369).

(93) *Ibid.* (p. 374, which quotes the report by Henry Cooke of the Commercial Intelligence Committee of the Board of Trade, (?)1899).

(94) JOHNSON, H. F. "Development of ice-breaking vessels for the U.S. Coast Guard", *Transactions of the Society of Naval Architects and Marine Engineers*, New York, Vol. 54, 1946, p. 112–51. (Reference to p. 114–16.)

(95) KALESNIK, S. V. "Prazdnovaniye stoletiya Geograficheskogo obshchestva SSSR i Vtoroy vsesoyuznyy geograficheskiy s"yezd" [Celebration of the centenary of the Geographical Society of the U.S.S.R. and the Second All-Union Geographical Congress], *Izv.Vse.Geog.Ob.* Tom 79, No. 2, 1947, p. 105–14. (Reference to p. 112.)

(96) KANTOROVICH, V. *S Karskoy ekspeditsii po severnomu morskomu puti* [*With the Kara expedition along the Northern Sea Route*]. Moscow, Leningrad, Izdatel'stvo "Molodaya Gvardiya", 1930. 208 p. (Reference to p. 148.)

(97) KARBATOV, V. P. "Polyarnyye stantsii v 1935 godu" [Polar stations in 1935], *Sov.Ark.* No. 4, 1935, p. 6–18.

(98) KARELIN, D. B. *More Laptevykh* [*Laptev Sea*]. Moscow, Leningrad, Izdatel'stvo Glavsevmorputi, 1947. 200 p.

(99) *Ibid.* (p. 64).

(100) *Ibid.* (p. 177).

(101) KARELIN, D. B. "Povysim kachestvo ledovykh prognozov" [Let us raise the quality of ice forecasts], *Sov.Ark.* No. 8, 1940, p. 25–29.

(102) KARELIN, D. B. "Problema kratkosrochnykh ledovykh prognozov" [The problem of short-term ice forecasts], *Sov.Ark.* No. 7, 1940, p. 10–17.

(103) KARELIN, D. B., and OVCHINNIKOV, I. G. "Kratkosrochnyye ledovyye prognozy" [Short-term ice forecasts], *Prob.Ark.* No. 3, 1940, p. 55–68. (Reference to p. 55–56.)

(104) *Ibid.* (p. 56).

(105) "Karskaya ekspeditsiya" [The Kara expedition], *Sib.Sov.Ent.* Tom. 2, 1931, cols. 543–49.

(106) "Karskaya ekspeditsiya" [The Kara expedition], *Byull.Ark. Inst.* No. 11/12, 1934, p. 410.

(107) "Karskoye more" [The Kara Sea], *Bol.Sov.Ent.* Tom 31, 1937, cols. 623–31.

(108) KHMYZNIKOV, P. K. "Plavaniye sudna *Belukha* v 1931 godu" [The voyage of the ship *Belukha* in 1931], *Byull.Ark.Inst.* No. 2, 1932, p. 24–27.

(109) KILESSO, A. "O novykh tipakh arkticheskikh sudov" [On new types of arctic ship], *Sov.Ark.* No. 6, 1940, p. 17–24.

(109a) KIZEVETTER, I. V. "Uluchshit' kachestvo soli, postupayushchey dlya rybnoy promyshlennosti·Dal'nego Vostoka" [Improve the quality of the salt arriving for the fishing industry of the Far East], *Rybnoye Khozyaystvo* [*Fisheries*], Moscow, No. 6, 1951, p. 23–27.

(110) KNIPOVICH, N. M. "Gidrologicheskaya s"yemka Barentsova morya" [Hydrological survey of the Barents Sea], *Byull.Ark.Inst.* No. 12, 1935, p. 440–42.

(111) KOLESOV, G. G., and POTAPOV, S. G. *Sovetskaya Yakutiya* [*Soviet Yakutiya*]. Moscow, Gosudarstvennoye Sotsial'no-Ekonomicheskoye Izdatel'stvo, 1937. 340 p. (Reference to p. 251.)

(112) KOMOV, N. N. "Meteo dlya meteo?" [Meteorology for the sake of meteorology?] *Sov.Ark.* No. 3, 1938, p. 11–13.

(113) KOMOV, N. N. "O rabote gidrometeorologicheskoy sluzhby Glavsevmorputi" [On the work of the hydrological and meteorological service of Glavsevmorput'], *Sov.Ark.* No. 7, 1939, p. 47–50.

(114) KONDAKOV, K. "Malokabotazhnyye perevozki po severnomu morskomu puti" [Short-range coastal freighting on the Northern Sea Route], *Sov.Ark.* No. 8, 1940, p. 17-25.

(115) KORNILOV, I. "Vodnyye puti Yakutskogo severa" [Waterways of the Yakutsk north], *Sov.Ark.* No. 7, 1937, p. 50-53.

(116) KORNILYUK, YU. N., KOCHETKOV, T. P., and YEMELYANTSEV, T. M. "Nordvik-Khatangskiy neftenosnyy rayon" [The Nordvik-Khatanga oil-bearing region], *Nedra Ark.* No. 1, 1946, p. 15-73. (Reference to p. 15-17.)

(117) KOROVKIN, I. P. "Materialy po gidrologii reki Khatangi" [Material on the hydrology of the river Khatanga], *Sev.Mor.Put'*, No. 16, 1940, p. 79-98.

(118) KOSOY, A. I. *Na vostochnom poberezh'ye Taymyrskogo poluostrova [On the east coast of Poluostrov Taymyr].* Moscow, Leningrad, Izdatel'stvo Glavsevmorputi, 1944. 159 p.

(119) KOSTYUK, A. G. "Arkticheskiye porty" [Arctic ports], in ZUBOV *et al.*, No. 405 below, p. 175-84.

(120) KOSTYUK, A. G. "O rabote porta Tiksi" [On the work of the port of Tiksi], *Sov.Ark.* No. 11, 1940, p. 26-32.

(121) KOSTYUK, A. G. "Port Provideniya", *Sov.Ark.* No. 4, 1940, p. 44-49.

(122) KOZLOV, M. I. "Tumany vdol' trassy severnogo morskogo puti" [Fogs along the course of the Northern Sea Route], *Trudy Ark.Inst.* Tom 109, 1937, p. 1-85. (Reference to p. 69-75.)

(123) KRASTIN, E. F. "Itogi i perspektivy" [Results and prospects], *Sov.Ark.* No. 2, 1936, p. 56-62.

(124) KRASTIN, E. F. "Severnyy morskoy put' v eksploatatsii" [The Northern Sea Route in use], *Sov.Ark.* No. 1, 1935, p. 20-23.

(125) KUBLITSKIY, G. *Yenisey reka sibirskaya [The Yenisey is a Siberian river].* Moscow, Leningrad, Gosudarstvennoye Izdatel'stvo Detskoy Literatury Ministerstva Pros-veshcheniya RSFSR, 1949. 288 p. (Reference to p. 281-84.)

(126) KURENKOV, A. "Kak rabotal rechnoy transport v 1938 godu i podgotovka k novoy navigatsii" [How river transport worked in 1938 and preparations for the new navigation season], *Sov.Ark.* No. 3, 1939, p. 65-68.

(127) KURENKOV, A. "Kolymo-Indigirskiy rechnoy transport" [Kolyma-Indigirka river transport], *Sov.Ark.* No. 10, 1939, p. 42-45.

(128) KURENKOV, A. "Reka Indigirka i yeye osvoyeniye" [The river Indigirka and the conquest of it], *Sov.Ark.* No. 7, 1940, p. 36-38.

(129) KUSOV, N. I., and LAPPO, V. I. "Mestorozhdeniye uglya Yurung-Tumus" [The Yurung-Tumus coal deposit], *Nedra Ark.* No. 2, 1947, p. 100-12. (Reference to p. 109.)

(130) LAKTIONOV, A. F. "Itogi issledovaniy ledyanogo pokrova morey sovetskoy arktiki i ledovyye prognozy" [Results of study of the ice cover of Soviet Arctic seas and ice forecasts], *Izv.Vse.Geog.Ob.* Tom 77, No. 6, 1945, p. 341-50. (Contains selected list of works on this subject published between 1920 and 1945. English summary of this paper appears in *Polar Record*, Vol. 5, No. 39, 1950, p. 468-73.)

(131) LAKTIONOV, A. F. "Nauchnyye rezul'taty arkticheskoy ekspeditsii na *Lomonosove* v 1931 godu. Gidrologiya i meteorologiya" [Scientific results of the arctic expedition in the *Lomonosov* in 1931. Hydrology and meteorology], *Trudy Ark.Inst.* Tom 18, 1935, p. 1-107.

(132) LAKTIONOV, A. F. *Severnaya Zemlya.* Moscow, Leningrad, Izdatel'stvo Glavsevmorputi, 1946. 152 p. (Reference to p. 126.)

(133) LAKTIONOV, A. F. "Zadachi ledovykh i gidrologicheskikh issledovaniy v arktike" [The tasks of ice and hydrological studies in the Arctic], *Prob.Ark.* No. 6, 1939, p. 5-10.

(134) LAKTIONOV, A. F., and DREMLYUG, V. V. "Analiz skorosti dvizheniya sudov vo l'dakh arkticheskikh morey" [Analysis of the speed of ships moving in ice in the arctic seas], *Prob.Ark.* No. 1, 1944, p. 5-32.

(135) LAPPO, S. D. *Spravochnaya knizhka polyarnika [Polar worker's handbook].* Moscow, Leningrad, Izdatel'stvo Glavsevmorputi, 1945. 423 p. (Reference to p. 274.)

(136) LAPPO, V. I. "Neftyanoye mestorozhdeniye Nordvik" [The Nordvik oil deposit], *Nedra Ark.* No. 1, 1946, p. 74-129. (Reference to p. 75.)

(137) *Ibid.* (p. 119).
(138) *Ibid.* (p. 126).
(139) LAPPO, V. I., and KUSOV, N. I. "Nordvikskoye mestorozhdeniye kamennoy soli" [The Nordvik deposit of rock salt], *Nedra Ark.* No. 2, 1947, p. 147–76. (Reference to p. 151.)
(140) LAVROV, A. "Taymyrskaya gidrograficheskaya ekspeditsiya 1932 goda na g/s *Taymyr*" [Taymyr hydrographic expedition of 1932 in the hydrographic ship *Taymyr*], *Byull.Ark.Inst.* No. 11/12, 1932, p. 253–56.
(141) LAVROV, B. V. "Itogi 1930 g. na Sibirskom severe" [Results of 1930 in the Siberian north], *Sov.Aziya*, No. 5/6, 1931, p. 53–58.
(142) LAVROV, B. V., and SHADRIN, N. YE. "Ekonomika gruzooborota severnogo morskogo puti" [The economics of freight turnover on the Northern Sea Route], *Sov.Ark.* No. 4, 1936, p. 11–24.
(143) *Law on the five-year plan for the rehabilitation and development of the national economy of the U.S.S.R.* London, Press Department of the Soviet Embassy, 1946. 32 p. (Reference to p. 7.)
(144) *Ibid.* (p. 16).
(145) LE-MYUR, V. "Snabzheniye severa Yakutii i Chukotki" [Supplying the north of Yakutiya and Chukotka], *Sov.Ark.* No. 2, 1935, p. 40–44.
(146) "Ledovyy patrul' v Barentsovom more" [Ice patrol in the Barents Sea], *Sov.Ark.* No. 9, 1939, p. 75–76.
(147) "Lenskaya morskaya transportnaya ekspeditsiya 1933 goda" [Lena sea transport expedition of 1933], *Byull.Ark.Inst.* No. 5, 1934, p. 216–18.
(148) LESGAFT, E. *L'dy severnago ledovitogo okeana i morskoy put' iz Yevropy v Sibir'* [*The ice of the Arctic Ocean and the sea route from Europe to Siberia*]. St Petersburg, O. N. Popov, 1913. 239 p.
(149) *Ibid.* (p. 15).
(150) LESLIE, A. *The arctic voyages of Adolf Erik Nordenskiöld, 1858–1879.* London, Macmillan, 1879. xvi+440 p.
(151) LEVICHEV, T. A. "Reki severa—na sluzhbu sotsialisticheskomu stroitel'stvu" [The rivers of the north should be put to the service of socialist construction], *Sov.Ark.* No. 9, 1937, p. 19–20.
(152) LEVONEVSKIY, D. A., comp. *S. O. Makarov i zavoyevaniye arktiki* [*S. O. Makarov and the conquest of the Arctic*]. Moscow, Leningrad, Izdatel'stvo Glavsevmorputi, 1943. 331 p. (Contains on p. 43–192 "*Yermak*" *vo l'dakh*, which is Makarov's own account of the early voyages of the *Yermak*.) (Reference to p. 9–40.)
(153) *Ibid.* (p. 48).
(154) *Ibid.* (p. 75).
(155) LIBMAN, A. YU. "Severnyy morskoy put" [The Northern Sea Route], *Za Ind.Sov. Vostoka*, No. 2, 1933, p. 65–80.
(156) LIED, JONAS. *Return to happiness.* London, Macmillan, 1943. xii+318 p.
(157) "List of synoptic stations of the U.S.S.R." *Publications du Secrétariat de l'Organisation Météorologique Internationale*, Lausanne, No. 9, Fasc. 2, Supplement 6, Annex, April 1949.
(158) *Lloyd's register of shipping.* London, Lloyd's, 1945.
(159) LOMAKIN, N. "Morskiye polyarnyye stantsii na 1 yanvarya 1938 g." [Coastal polar stations on 1 January 1938], *Prov.Ark.* No. 2, 1938, p. 208–11.
(160) *Lotsiya morya Laptevykh,* 1938. *Dopolneniye No. 2,* 1940. *Ispravlennoye na 1 yanvarya 1940 g.* [*Laptev Sea pilot*, 1938. *Supplement No. 2*, 1940. *Corrected to 1 January 1940*]. Leningrad, Izdatel'stvo Gidrograficheskogo Upravleniya VMF, 1940.
(161) *Lotsiya Vostochnosibirskogo morya* [*East Siberian Sea pilot*]. Leningrad, Izdaniye Gidrograficheskogo Upravleniya Glavsermorputi, 1939. (Reference to p. 84.)
(162) LUR'YE, K. M. "Amderminskiy plavikovyy shpat" [Amderma fluorspar], *Sov.Ark.* No. 10, 1937, p. 65–68.
(163) LYUBARSKIY, I. "Perspektivy Pridivnenskoy sudoverfi" [Prospects of the Pridivnensk shipyard], *Sov.Ark.* No. 5, 1938, p. 114–17.
(164) MARGOLIN, A. "Anglo-frantsuskaya interventsiya na severe Rossii i severnyy morskoy put" [Anglo-French intervention in the north of Russia and the Northern Sea Route], *Sov.Ark.* No. 11, 1940, p. 77–80.

(165) *Ibid.* (p. 80).

(166) MARGOLIN, A. "Igarke—desyat' let" [Igarka is ten years old], *Sov.Ark.* No. 11, 1939, p. 91–102.

(167) MARGOLIN, A. "Puti zavoza gruzov na krayniy sever" [Freight routes to the far north], *Sov.Ark.* No. 6, 1939, p. 32–46.

(168) MASON, KENNETH. "Notes on the Northern Sea Route", *Geog.Journal*, Vol. 96, No. 1, 1940, p. 27–41.

(169) "Materialy po klimatologii polyarnykh oblastey SSSR" [Material on the climatology of the polar regions of the U.S.S.R.]. Published as volumes of *Trudy Ark.Inst.* (Based on polar stations' observations, for which estimates of reliability are given.)

(170) MATTERS, LEONARD. *Through the Kara Sea.* London, Skeffington, 1932. 284 p.

(171) MATUSEVICH, N. K., and KOGANOV, S. M. "Novyye sovetskiye ledokoly" [New Soviet icebreakers], *Sov.Ark.* No. 9, 1937, p. 83–84.

(172) MAZO, A. "Odna iz vazhneyshikh khozyaystvenno-politicheskikh zadach" [One of the most important economic and political tasks], *Sov.Ark.* No. 12, 1938, p. 43–48.

(173) McMURTRIE, F. E., ed. *Jane's Fighting Ships*, 1946–47. London, Sampson Low, Marston, 1947. 61 adv., xxiii, A 21, 471 p. (Reference to p. 437.)

(174) McMURTRIE, F. E., ed. *Jane's Fighting Ships*, 1947–48. London, Sampson Low, Marston, 1948. 56 adv., xxiv, A 27, 498 p.

(175) *Ibid.* (p. 287).

(176) *Ibid.* (p. 288–89).

(177) MIKHAYLOV, A. P. "Set' polyarnykh stantsiy v tret'yey pyatiletke" [The network of polar stations in the third five-year plan]. *Sov.Ark.* No. 10, 1937, p. 25–28.

(178) MINEYEV, A. I. "O prokhode v Yeniseyskiy zaliv prolivom Ovtsyna" [On entering Yeniseyskiy Zaliv by way of Proliv Ovtsyna], *Sov.Ark.* No. 5, 1941, p. 21.

(179) MINEYEV, A. I. "Yeniseyskaya operatsiya" [The Yenisey operation], in ZUBOV, *et al.*, No. 405 below, p. 103–49.

(180) *Ibid.* (p. 136–37).

(181) MISYUREV, V. "Peleduyskaya sudoverf' " [The shipyard at Peleduy], *Sov.Ark.* No. 3, 1940, p. 86–87.

(182) MOISEYEV, I. V., and TEBEN'KOV, V. P. "Geologicheskoye stroyeniye i poleznyye iskopayemyye vodorazdela rek Nizhney Tunguski, Sukhoy Tunguski i Bakhty" [Geological structure and useful minerals of the watershed of the rivers Nizhnyaya Tunguska, Sukhaya Tunguska and Bakhta], *Trudy Ark.Inst.* Tom 139, 1939, p. 7–108. (Reference to p. 97.)

(183) MOKRINSKIY, V. V., and PONOMAREV, T. N. "Uglenosnost' sovetskoy arktiki" [Coal-bearing potentialities of the Soviet Arctic], *Prob.Ark.* No. 7/8, 1939, p. 5–40.

(184) *Ibid.* (p. 7).

(185) *Ibid.* (p. 11).

(186) *Ibid.* (p. 32–33).

(187) *Ibid.* (p. 36).

(188) MOLDAVSKIY, M. L., and ROKHLIN, M. I. "Raboty Chaunskoy geologopoiskovoy ekspeditsii 1936–37 g." [Work of the Chaun geological prospecting expedition of 1936–37], *Prob.Ark.* No. 1, 1938, p. 102–09.

(189) MOLODETSKIY, K. G. "Kamennougol'nyye bazy severnogo morskogo puti" [Sources of coal on the Northern Sea Route], *Izv.Vse.Geog.Ob.* Tom 73, No. 1, 1941, p. 113–17.

(190) MOLODYKH, I. F. "Kak osvoit' Kolymsko-Indigirskiy kray" [How to conquer the Kolyma-Indigirka region], *Sov.Aziya*, No. 9/10, 1931, p. 57–75.

(191) MONASTYRSKIY, A. S. "Ugol'nyye resursy" [Coal resources], *Sov.Ark.* No. 7, 1937, p. 67–68.

(192) MORA, SYLWESTER, and ZWERNIAK, PIOTR. *Sprawiedliwość sowiecka* [*Soviet justice*]. Italy, 1945. 275 p. (Reference to p. 87–89.)

(193) Moscow Radio, 16 July 1946.

(194) Moscow Radio, 18 February 1947.

(195) Mutafi, N. N. "K geologii i uglenosnosti zapadnogo Taymyra po materialam dvukh peresecheniy 1938 i 1939 gg." [Geology and coal-bearing properties of western Taymyr on the basis of two crossings made in 1938 and 1939], *Prob.Ark.* No. 5, 1940, p. 35–55.

(196) "Na priyeme v Kremle" [At the reception in the Kremlin], *Sov.Ark.* No. 3, 1936, p. 6–27. (Reference to p. 25.)

(197) "Na stupen' vyshe" [A step higher], *Sov.Ark.* No. 5, 1935, p. 3–6. (Reference to p. 3.)

(198) "Na Tobol'skoy sudoverfi" [At the Tobol'sk shipyard], *Sov.Ark.* No. 4, 1938, p. 64–65.

(199) Nansen, Fridtjof. *Farthest north.* London, George Newnes, 1898. 2 vols.

(200) Nansen, Fridtjof. *Through Siberia, the land of the future.* London, William Heinemann, 1914. xvi + 478 p.

(201) *Ibid.* (p. 156).

(202) *Ibid.* (p. 446).

(203) *Ibid.* (p. 447).

(204) *Ibid.* (p. 448).

(205) Natsarenus, S. P. "Takova li ekonomika gruzooborota?" [Are the economics of freight turnover like that?], *Sov.Ark.* No. 10, 1936, p. 25–28.

(206) "Nauchnyye rezul'taty ekspeditsii na *Malygine* na Zemlyu Frantsa-Iosifa v 1932 g. Gidrologiya i meterologiya" [Scientific results of the expedition in the *Malygin* to Zemlya Frantsa-Iosifa (Franz Josef Land) in 1932. Hydrology and meteorology], *Trudy Ark.Inst.* Tom 34, 1935, p. 1–52.

(207) Nazarov, B. "Na reke Yane" [On the river Yana], *Sov.Ark.* No. 11, 1940, p. 34–37.

(208) Nazarov, M. "Sotsialisticheskoye sorevnovaniye na polyarnykh stantsiyakh" [Socialist emulation at polar stations], *Sov.Ark.* No. 8, 1940, p. 66–68.

(209) Nazarov, V. "Rezul'taty sravnitel'nykh ispytaniy modeli ledokola novogo tipa" [Results of comparative tests of a model of a new-type icebreaker], *Mor.Flot*, No. 5/6, 1946, p. 12–15.

(210) Nazarov, V. S. "Sovremennoye sostoyaniye ledovykh prognozov" [The present position of ice forecasts], *Sov.Ark.* No. 2, 1938, p. 36–46.

(211) Nikolayev, V. "Severo-vostochnyy prokhod" [The North East Passage], *Izv.Arkh. Ob.iz.Russ.Sev.* No. 6, 1913, p. 241–44.

(212) Nikol'skiy, V. N. "Severnyy morskoy put' i nashi issledovaniya polyarnykh stran" [The Northern Sea Route and our exploration of polar countries], *Izv.Arkh.Ob.iz. Russ.Sev.* No. 19, 1914, p. 633–39. (Reference to p. 638.)

(213) Nordenskiöld, A. E. *Sur la possibilité de la navigation commerciale dans la mer glaciale de Sibérie.* Stockholm, Kongl. Boktryckeriet, 1879. 36 p. (Reference to p. 30, 35.)

(214) *Ibid.* (p. 34, 35).

(215) Nordenskiöld, A. E. *The voyage of the Vega round Asia and Europe.* London, Macmillan, 1881. 2 vols.

(216) "Norges Bergverksdrift, 1948." *Norges Ofisielle Statistikk,* Oslo, Rekke 11, No. 8, 1950. (Reference to p. 11.)

(217) *Ibid.* (p. 53).

(218) "Noril'skiy gornopromyshlennyy rayon" [Noril'sk industrial mining region], *Sib.Sov. Ent.* Tom 3, 1932, cols. 797–800.

(219) "The north Pechora railway and the development of the Pechora coalfields", *Polar Record*, Cambridge, Vol. 4, No. 29, 1945, p. 236–38.

(220) Notkin, A. I. "Severnyy morskoy put'" [The Northern Sea Route], *Sev.Aziya*, No. 1/2, 1925, p. 28–43; No. 4, 1925, p. 53–75. (Reference to p. 31, 36.)

(221) *Ibid.* (p. 54–56).

(222) "Novosti arktiki" [Arctic news], *Sov.Ark.* No. 8, 1940, p. 97.

(223) *Ibid.* (p. 98).

(224) "Novyy Port", *Sib.Sov.Ent.* Tom 3, 1932, cols. 792–94.

(225) "O rabote Glavsevmorputi za 1937 god" [On the work of Glavsevmorput' in 1937], *Sov.Ark.* No. 5, 1938, p. 21.

(226) "O razvitii severnogo morskogo puti" [On the development of the Northern Sea Route], *Sov.Sev.* No. 5, 1934, p. 110–11. (English translation appears in Taracouzio, No. 330 below, p. 389–91.)

(227) "Obzor plavaniy v vodakh arktiki v 1934 g." [Survey of voyages in arctic waters in 1934], *Byull.Ark.Inst.* No. 1/2, 1935, p. 1–4.

(228) "Organizovat' promezhutochnyye bunkernyye bazy" [Intermediate bunkering bases should be organized], *Sov.Ark.* No. 2, 1938, p. 113.

(229) "Organizatsiya Bukhtugol'stroya" [The organisation of Bukhtugol'stroy], *Sov.Ark.* No. 1, 1940, p. 98.

(230) ORLOV, A. "Arkticheskaya Komissiya" [Arctic Commission], *Sov.Sev.* No. 2, 1930, p. 111–16.

(231) OSTREKIN, M. YE. "Novyye magnitnyye i ionosfernyye stantsii v sovetskoy arktike" [New magnetic and ionospheric stations in the Soviet Arctic], *Prob.Ark.* No. 2, 1944, p. 120–21.

(232) OSTROUMOVA, V. P. "Delo osvoyeniya severa dvigat' vpered!" [Carry forward the conquest of the north!], *Sov.Ark.* No. 3, 1935, p. 11–12.

(233) PAPANIN, I. D. "Itogi 1940 goda i zadachi navigatsii 1941 goda" [The results of 1940 and the tasks for the navigation season of 1941], *Sov.Ark.* No. 4, 1941, p. 1–24.

(234) PAPANIN, I. D. "Na novyye uspekhi" [To new successes], in ZUBOV, *et al.*, No. 405 below, p. 3–10.

(235) PAPANIN, I. D. "Rech' tov. I. Papanina na XVIII s"yezde VKP(b)" [The speech of comrade I. Papanin at the 18th Congress of the All-Union Communist Party (bolsheviks)], *Sov.Ark.* No. 4, 1939, p. 89–94.

(236) *Ibid.* (p. 91).

(237) "Partiyno-khozyaystvennyy aktiv Glavsevmorputi" [Active party workers of Glavsevmorput'], *Sov.Ark.* No. 1, 1940, p. 18–29.

(238) "Penalties for disclosure of state secrets", *Soviet Monitor*, London, 10 June 1947.

(239) "Perebroska gruzov po severnomu morskomu puti" [Freighting on the Northern Sea Route], *Byull.Ark.Inst.* No. 9, 1935, p. 291.

(240) PETROV, M. "Ledokol'nyy flot SSSR" [The icebreaker fleet of the U.S.S.R.], *Mor. Flot*, No. 11, 1947, p. 37–41.

(241) "Po-bol'shevistski izuchat' arktiku" [Let us study the Arctic in a bolshevik way], *Sov.Ark.* No. 2, 1935, p. 3–7.

(242) "Po-bol'shevistski vypolnim resheniye Sovnarkoma SSSR ob uluchshenii raboty Glavsevmorputi" [Let us carry out in a bolshevik way the decision of the Council of People's Commissars of the U.S.S.R. on the improvement of the work of Glavsevmorput'], *Sov.Ark.* No. 9, 1938, p. 3–4.

(243) "Polyarnyye stantsii k novomu godu" [Polar stations in the new year], *Sov.Ark.* No. 12, 1937, p. 57–59.

(244) POPOV, V. L. "Tri magistrali" [Three trunk lines], *Sov.Aziya*, No. 3/4, 1930, p. 28–44. (Reference to p. 37.)

(245) "Postanovleniye Soveta narodnykh komissarov Soyuza SSR" [Decree of the Council of People's Commissars of the U.S.S.R.], *Sov.Ark.* No. 9, 1938, p. 5.

(246) "Pozornoye otstavaniye" [Shameful lagging], *Sov.Ark.* No. 10, 1937, p. 3–8.

(247) *Pravda*, Moscow, 14 October 1945.

(248) *Pravda*, Moscow, 3 December 1945.

(249) *Pravda*, Moscow, 11 March 1946.

(250) *Pravda*, Moscow, 27 March 1949.

(251) *Pravda*, Moscow, 9 and 10 September 1949.

(252) PRIK, Z. M. "Klimaticheskiy ocherk Karskogo morya" [Climatic outline of the Kara Sea], *Trudy Ark.Inst.* Tom 187, 1946, p. 5–442.

(253) "Pyatiletiye Gidrograficheskogo Instituta" [Fifth anniversary of the Hydrographic Institute (Gidrograficheskiy Institut)], *Prob.Ark.* No. 4, 1940, p. 106.

(254) "Rabota severo-Yakutskogo rechnogo parokhodstva [Work of the north Yakutiya river shipping company], *Sov.Ark.* No. 1, 1940, p. 97.

(255) *Radio weather aids. H.O. Publication No. 206.* Washington, U.S. Navy Department, 1946. (Amended to September 1948). (Reference to p. 6–16, 6–19, 6–20.)

(256) RAKITOV, A. I. "Migratsiya nefti v usloviyakh vechnoy merzloty" [Movement of oil in conditions of permanent frost], *Prob.Ark.* No. 6, 1940, p. 40–58. (Reference to p. 49–51.)

(257) *Ibid.* (p. 51–52).

(258) RAMSAY, HENRIK. *I Kamp med östersjöns isar.* Helsingfors, Holger Schildts Forlag, 1947. 415 p. (Reference to p. 53–54.)

(259) RATMANOV, G. YE. "Beringovaya partiya Tikhookeanskoy ekspeditsii Godudarstven-nogo Gidrologicheskogo Instituta" [Bering Strait party of the Pacific Ocean expedition of the State Hydrological Institute (Gosudarstvennyy Gidrologi-cheskiy Institut)], *Byull.Ark.Inst.* No. 1/2, 1933, p. 8–9.

(260) "Reorganizatsiya Arkticheskogo Instituta" [Reorganisation of the Arctic Institute], *Prob.Ark.* No. 7/8, 1940, p. 106–07.

(261) "Reshitel'no perestroit' rabotu" [Work should be decisively reorganised], *Sov.Ark.* No. 5, 1938, p. 22–24.

(262) RIKHTER, B. "Matochkin Shar", *Sov.Ark.* No. 5, 1938, p. 81–88.

(263) RODEVICH, V. M., ed. *Spravochnik po vodnym resursam SSSR. Tom XVI. Leno-Yeniseyskiy rayon* [*Handbook of the water resources of the U.S.S.R. Vol. 16. Lena-Yenisey region*]. Leningrad, Moscow, Redaktsionno-Izdatel'skiy Otdel TsUYeGMS, 1936, 1217 p., in 2 parts.

(264) *Ibid.* (p. 908).

(265) ROGATKO, G. "Osvoyeniye reki Anabara", [The conquest of the river Anabar], *Sov.Ark.* No. 8, 1939, p. 35–42.

(266) "Russia at war", *Economist*, London, Vol. 145, No. 5224, 9 October 1943, p. 497.

(267) RYAZANTSEVA, Z. "Yuzhno-Taymyrskiy vodnoy put' kak uchastok severnogo morskogo puti" [The south Taymyr waterway as part of the Northern Sea Route], *Prob.Ark.* No. 9, 1939, p. 89–92.

(268) RYBNIKOV, S. I. "Arkhangel'skaya verf' derevyannogo sudostroyeniya" [Ark-hangel'sk yard for building wooden ships], *Sov.Ark.* No. 7, 1936, p. 53–54.

(269) RYZHOV, V. "Snizit' sebestoimost' po gruzochno-razgruzochnykh rabot v portakh" [Lower costs on loading and unloading work in ports], *Sov.Ark.* No. 9, 1940, p. 18–21.

(270) SAMOYLOVICH, R. L. "Ekspeditsiya na l/p *Sadko* v 1936 godu" [The expedition in the icebreaking ship *Sadko* in 1936], *Byull.Ark.Inst.* No. 10/11, 1936, p. 457–59.

(271) SAMOYLOVICH, R. L. "Ekspeditsiya na ledokol'nom parokhode *Rusanov*" [The expedi-tion in the icebreaking ship *Rusanov*], *Byull.Ark.Inst.* No. 8/10, 1932, p. 190–93.

(272) SAMOYLOVICH, R. L. *Vo l'dakh arktiki. Pokhod "Krasina" letom 1928 goda. Tretye izdaniye* [*In the ice of the Arctic. The voyage of the "Krasin" in the summer of 1928. 3rd edition*]. Leningrad, Izdaniye Vsesoyuznogo Arkticheskogo Instituta, 1934. 340 p.

(273) SAMOYLOVICH, R. L. "Za pyatnadtsat' let" [For fifteen years], *Byull.Ark.Inst.* No. 3/4, 1935, p. 56–61.

(274) SAPTSOV, –. "Rabota Glavnogo Upravleniya Severnogo morskogo puti v vostochnom sektore arktiki v 1934 godu" [The work of Glavsevmorput' in the eastern sector of the Arctic in 1934], *Byull.Ark.Inst.* No. 11/12, 1934, p. 398.

(275) SEEBOHM, HENRY. *The birds of Siberia*. London, John Murray, 1901. xix + 512 p. (Reference to p. 263.)

(276) SERGEYEVSKIY, –. "Sovremennyye karty polyarnogo morya u beregov Sibiri" [Present charts of the polar sea in Siberian offshore waters], *Sev.Aziya*, No. 4, 1928, p. 100–08. (Reference to p. 102.)

(277) SERGIYEVSKIY, D. N. "S karavanom zemlecherpalok po severnomu morskomu puti" [With a convoy of dredgers along the Northern Sea Route], in ZUBOV, *et al.*, No. 405 below, p. 95–102.

(278) "Severnyy morskoy put" [The Northern Sea Route], *Bol.Sov.Ent.* Tom 50, 1944, cols. 579–91.

(279) SHIBINSKIY, V. G. "Karskaya morskaya operatsiya 1932 goda" [Kara Sea operation of 1932], *Byull.Ark.Inst.* No. 4, 1933, p. 77–80.

(280) SHIBINSKIY, V. G. "Karskaya operatsiya 1933 goda" [Kara operation of 1933], *Byull.Ark.Inst.* No. 5/6, 1935, p. 139–41.

(281) SHIBINSKIY, V. G. "Karskaya operatsiya 1934 goda" [Kara operation of 1934], *Byull.Ark.Inst.* No. 5/6, 1935, p. 141–45.

(282) SHIBINSKIY, V. G. "Karskaya morskaya operatsiya 1935 goda" [Kara Sea operation of 1935], *Byull.Ark.Inst.* No. 3, 1936, p. 115–17.

(283) SHIMANSKIY, Yu. A. "Uslovnyye izmeriteli ledovykh kachestv sudna" [Conditional standards for measuring ice qualities of ships], *Trudy Ark.Inst.* Tom 130, 1938, p. 3–59.

(284) SHIRSKOV, P. P. "Novyy etap raboty Sevmorputi" [A new phase in the work of the Northern Sea Route], *Sov.Ark.* No. 12, 1940, p. 3–7.

(285) SHIRSHOV, P. P. "Obespechim uspekh arkticheskoy navigatsii 1939 goda" [Let us secure the success of the arctic navigation season of 1939], *Sov.Ark.* No. 7, 1939, p. 3–7.

(286) SHISHKOVA, L. "Kratkiye svedeniya o poleznykh iskopayemykh sovetskogo sektora arktiki" [Brief account of the useful minerals of the Soviet sector of the Arctic], *Byull.Ark.Inst.* No. 10, 1934, p. 370–73. (English translation in TARACOUZIO, No. 330 below, p. 463–65.)

(287) SHMIDT, O. YU. "Arktika v 1935 godu" [The Arctic in 1935], *Byull. Ark.Inst.* No. 5/6, 1935, p. 123–35.

(288) SHMIDT, O. YU. "The *Chelyuskin* expedition", trans. in BROWN, ALEC, *The voyage of the "Chelyuskin"*, London, Chatto and Windus, 1935, p. 1–22.

(289) SHMIDT, O. YU. *Osvoyeniye severnogo morskogo puti i zadachi sel'skogo khozyaystva kraynego severa* [*The conquest of the Northern Sea Route and the tasks of agriculture in the far north*]. Moscow, Leningrad, Izdatel'stvo Glavsevmorputi, 1937. 15 p. (Reference to p. 1.)

(290) SHMIDT, O. YU. "Polyarnaya magistral" [Polar trunk line], *Byull. Ark. Inst.* No. 10/11, 1936, p. 441–45.

(291) SHMIDT, O. YU., and BERGAVINOV, S. A. "Arkticheskaya navigatsiya okonchena" [The arctic navigation season has ended], *Byull.Ark.Inst.* No. 10/11, 1936, p. 440–41.

(292) SHMIDT, O. YU., and BERGAVINOV, S. A. "Severnyy morskoy put' prevrashchayetsya v normal'no deystvuyushchiy put'" [The Northern Sea Route is becoming a normally working route], *Byull.Ark.Inst.* No. 11, 1935, p. 371–72.

(293) SHOKAL'SKIY, YU. M. "Les recherches des Russes de la route maritime de Sibérie", *Report of the Sixth International Geographical Congress*, London, 1896, p. 239–46.

(294) SHOKAL'SKIY, YU, M. "Russian navigators in the Arctic Ocean in 1895–96", *Geog. Journal*, Vol. 12, 1898, p. 172–76.

(295) SHOKAL'SKIY, YU. M. "A short account of the Russian Hydrographical Survey", *Geog. Journal*, Vol. 29, 1907, p. 626–49.

(296) SHUL'TS, P. I. "Zyryanskiy kamennougol'nyy rayon" [Zyryanka coal-bearing region], *Prob.Ark.* No. 7/8, 1939, p. 70–74.

(297) SHUSTROV, M. "Avangardnaya rol' kommunistov na Kachugskoy sudoverfi" [Leading part played by communists at the Kachuga shipyard], *Sov. Ark.* No. 2, 1940, p. 78–80.

(298) SIBIRYAKOV, A. M. *Zur Frage von den äusseren Verbindungen Sibiriens mit Europa.* Zürich, Aktien Buchdrückerei Zürich, 1910, 76 p. (Reference to p. 9.)

(299) *Ibid.* (p. 11).

(300) *Ibid.* (p. 15).

(301) SIDOROV, M. K. *Sever Rossii* [*The north of Russia*]. St Petersburg, Pochtovyy Departament, 1870. 557 p. (Reference to p. 76, where Kruzenshtern's remark is quoted.)

(302) SIDOROV, M. K. *Trudy dlya oznakomleniya s severom Rossii* [*Work in publicising the north of Russia*]. St. Petersburg, D. I. Shemetkin, 1882. 342 p. (Reference to first (unnumbered) page of the introduction.)

(303) SLAVIN, S. "Glavsevmorput' v tret'yey stalinskoy pyatiletke" [Glavsevmorput' in the third Stalin five-year plan], *Sov.Ark.* No. 5, 1939, p. 25–37.

(304) SMIRNOV, G. S. "Nauchno-issledovatel'skuyu rabotu nado organizovat" [Scientific research work must be organised], *Sov.Ark.* No. 7, 1936, p. 101.

(305) SMOLKA, H. P. *Forty thousand against the Arctic.* London, Hutchinson, 1937, 288 p. (Reference to p. 146–62.)

(306) *Sobraniye postanovleniy i rasporyazheniy pravitel'stva SSSR* [*Collected decrees and ordinances of the Government of the U.S.S.R.*], Moscow. (Reference to 1938, § 184.)

(307) *Ibid.* (1939, § 96–97).

(308) *Ibid.* (1940, § 442).

(309) *Ibid.* (1941, § 76. English translation at Appendix VI).

(310) *Ibid.* (1945, § 135).

(311) *Sobraniye postanovleniy i rasporyazheniy soveta ministrov SSSR* [*Collected decrees and ordinances of the council of ministers of the U.S.S.R.*], Moscow. (Reference to 1946, § 214.)

(312) *Sobraniye zakonov i rasporyazheniy pravitel'stva SSSR* [*Collected laws and ordinances of the government of the U.S.S.R.*], Moscow. (Reference to 1932, Part I, § 522.)*

(313) *Ibid.* (1932, Part II, § 262).

(314) *Ibid.* (1933, Part I, § 124).*

(315) *Ibid.* (1933, Part I, § 265).

(316) *Ibid.* (1935, Part I, § 59)*

(317) *Ibid.* (1936, Part I, § 317).*

(318) SOKOLOV, A. A. "Zadachi gidrograficheskikh issledovaniy v Sibirskom more" [Tasks of hydrographic study in the Siberian Sea], *Sev.Aziya*, No. 3, 1928, p. 70–78.

(319) *Sov.Ark.*, No. 10/11, 1938. (This issue contains a number of articles on the activities of Komsomol in the Arctic.)

(320) *Soviet News*, London, 12 September 1946.

(321) *Soviet News*, London, 4 December 1946.

(322) STALIN, I. V. *Problems of Leninism, 11th edition.* Moscow, Foreign Languages Publishing House, 1947. 692 p. (Reference to p. 411.)

(323) "Standartnyye gidrologicheskiye razrezy v severnykh moryakh" [Standard hydrological sections in northern seas], *Prob.Ark.* No. 5, 1940, p. 98–100.

(324) STAROKADOMSKIY, L. "Plavaniya na ledokolakh *Taymyr* i *Vaygach* (1910–15 gg.)" [Voyages in the icebreakers *Taymyr* and *Vaygach* (1910–15)], *Sov.Ark.* No. 8, 1940. p. 69–81.

(325) STAVNITSER, M. *Russkiye na Shpitsbergene* [*Russians in Spitsbergen*], Moscow, Leningrad, Izdatel'stvo Glavsevmorputi, 1948. 150 p. (Reference to p. 73, quoting *Pravda* of 15 January 1947.)

(326) *Ibid.* (p. 88–90).

(327) STEPANOV, N. P. "Perevozki po severnomu morskomu puti" [Freightage on the Northern Sea Route], *Sov.Ark.* No. 12, 1936, p. 79–83.

(328) "Stroitel'stvo sudoremontnogo zavoda Glavsevmorputi" [Construction of the ship repair yard for Glavsevmorput'], *Prob.Ark.* No. 7/8, 1939, p. 107–08.

(329) SVERDRUP, H. U. "General report of the expedition", *The Norwegian North Polar expedition with the "Maud", 1918–25. Scientific results*, Bergen, Vol. 1*a*, No. 1, 1933, p. 1–22.

(330) TARACOUZIO, T. A. *Soviets in the Arctic.* New York, Macmillan, 1938. xvi+563 p.

(331) *Ibid.* (p. 73).

(332) TARSHIS, M. K. "Nauchnyye itogi issledovaniya prochnosti ledokol'nykh sudov" [Scientific results of the investigation of the resistance of icebreakers], *Prob.Ark.* No. 2, 1938, p. 115–43.

(333) TEBEN'KOV, V. P. "K poiskam prigodnykh dlya flota ugley na vostochnom poberezh'ye Yeniseyskogo zaliva (Zapadnyy Taymyr)" [Coal suitable for fleet use found on the eastern shore of Yeniseyskiy Zaliv (western Taymyr)], *Prob.Ark.* No. 12, 1939, p. 57–62.

(334) TEBEN'KOV, V. P. "Rezul'taty opytnogo szhiganiya uglya iz rayona reki Krest'yanki (Zap. Taymyr) na p/kh *Arkos*" [Results of experimental burning of coal from the region of the river Krest'yanka (western Taymyr) in the steamer *Arkos*], *Prob.Ark.* No. 12, 1940, p. 67–71.

(335) TEBEN'KOV, V. P. "Taymyrskiy uglenosnyy basseyn" [Taymyr coal basin], *Prob.Ark.* No. 2, 1939, p. 74–80.

(336) TELEGIN, A. "Sovremennoye sostoyaniye elektrodvizheniya sudov i perspektivy ispol'zovaniya gazoturboelektricheskoy peredachi na sudakh ledovogo plavaniya" [The present position with regard to electric motive power for ships and the prospects for using gas turbo-electric transmission in ships designed for ice navigation], *Mor.Flot*, No. 4, 1947, p. 5–9.

(337) TEL'NOV, I. "Problema skorostey dvizheniya transportnogo flota'" [The problem of speed in a transport fleet], *Sov.Ark.* No. 4, 1941, p. 41–49.

* English translations of these items are found in No. 330 below, p. 383–87, 393–400.

152 REFERENCES

(338) Tikhomirov, Ye. I. "Meteorologicheskiye issledovaniya v sovetskoy arktike za 25 let (1920–45)" [Meteorological studies in the Soviet Arctic during 25 years (1920–45)], *Izv.Vse.Geog.Ob.* Tom 77, No. 6, 1945, p. 322–27.

(339) Timofeyev, V. T. "Pervyy reys e/s *Akademik Shokal'skiy*" [The first trip of the expedition ship *Akademik Shokal'skiy*], *Prob.Ark.* No. 7/8, 1939, p. 108–09.

(340) Toll, Emmy von, ed. *Die russische Polarfahrt der "Sarja"*, 1900–02. Berlin, Georg Reimer, 1909. vi + 635 p.

(341) "Torgovyy reys iz ust'ya Leny v ust'ye Yany" [Freight voyage from the mouth of the Lena to the mouth of the Yana], *Byull.Ark.Inst.* No. 8, 1933, p. 236.

(342) "Tretiy pyatiletniy plan razvitiya narodnogo khozyaystva SSSR (1938–42 gg.). Rezolyutsiya XVIII s"yezda VKP(b) po dokladu tov. V. Molotova" [Third five-year plan for the development of the national economy of the U.S.S.R. (1938–42). Resolution of the 18th Congress of the All-Union Communist Party (bolsheviks) on the report of comrade V. Molotov], *Sov.Ark.* No. 4, 1939, p. 38–60.

(343) Trofimov, I. "V otryve ot osnovnykh zadach" [Out of touch with basic tasks], *Sov.Ark.* No. 7, 1938, p. 23–28.

(344) *Trudy dreyfuyushchey stantsii "Severnyy Polyus"* [*Transactions of the "North Pole" drifting station*]. Moscow, Leningrad, Izdatel'stvo Glavsevmorputi, 1940–45. 2 vols. (Only two vols. out of the intended four have so far (1950) appeared in western Europe.)

(345) Tsvetkova, A. N., ed. *Vladimir Aleksandrovich Rusanov*. Moscow, Leningrad, 1945. 428 p. (Reference to p. 50 ff.)

(346) "Ugol'naya baza na ov-e Diksona" [Coal base on Ostrov Diksona], *Byull.Ark.Inst.* No. 10, 1934, p. 376.

(347) "Ukaz prezidiuma verkhnogo soveta SSSR o nagrazhdenii rabotnikov Glavsevmorputi" [Decree of the Presidium of the Supreme Soviet of the U.S.S.R. on rewarding Glavsevmorput' workers], *Sov.Ark.* No. 5, 1940. p. 3–14.

(348) "Uluchshim rabotu polyarnoy radioseti" [Let us improve the work of the polar radio network], *Sov.Ark.* No. 3, 1938, p. 15.

(349) Ushakov, G. A. "Pamyati S. S. Kameneva, 1881–1932" [To the memory of S. S. Kamenev, 1881–1932], *Sov.Ark.* No. 10, 1936, p. 11–12.

(350) "V Glavnom Upravlenii Sevmorputi" [In the Chief Administration of the Northern Sea Route], *Sov.Ark.* No. 1, 1938, p. 98–99.

(351) Vasil'yev, N. *Karskaya ekspeditsiya* [*The Kara expedition*], Moscow, Redaktsiya Izdaniy NKVT, 1921. 44 p.

(352) Vavul, A., and Rapoport, I. "Rastvoreniye bogkheda Olenekskogo mestorozhdeniya" [Solution of the boghead coal from the Olenek deposit], *Nedra Ark.* No. 2, 1947, p. 138–46.

(353) Vaysutov, N. "Polveka nazad, kogda 'moglo i ne byt' gol'fshtrema'" [Half a century ago, when "perhaps the Gulf-stream did not exist"], *Sev.Aziya*, No. 1, 1928, p. 64–74. (Reference to p. 66.)

(354) *Ibid.* (p. 72).

(355) *Ibid.* (p. 72–73).

(356) Vinogradov, I. V. *Suda ledovogo plavaniya* [*Ships for ice navigation*]. Moscow, Glavnaya Redaktsiya Literatury po Sudostroyeniyu, 1946. 239 p. (Reference to p. 14).

(357) *Ibid.* (p. 16).

(358) *Ibid.* (p. 17–25).

(359) *Ibid.* (p. 19).

(360) *Ibid.* (p. 22–23; 26–33).

(361) *Ibid.* (p. 25–26).

(362) *Ibid.* (p. 77).

(363) *Ibid.* (p. 107).

(364) *Ibid.* (p. 113–16 and elsewhere, *passim*).

(365) Vitel's, L. "Karskaya morskaya ekspeditsiya 1930 goda" [Kara Sea operation of 1930], *Byull.Ark.Inst.* No. 3/4, 1931, p. 43–45.

(366) Vize, V. Yu. *Klimat morey sovetskoy arktiki* [*The climate of the seas of the Soviet Arctic*]. Moscow, Leningrad, Izdatel'stvo Glavsevmorputi, 1940. 124 p. (Reference to p. 45, 48.)

(367) *Ibid.* (p. 121).

(368) VIZE, V. YU. "Ledovyye prognozy dlya arkticheskikh morey" [Ice forecasts for arctic seas], *Sov.Ark.* No. 3, 1935, p. 25–30.

(369) VIZE, V. YU. *Morya sovetskoy arktiki* [*Seas of the Soviet Arctic*]. Leningrad, Izdatel'stvo Glavsevmorputi, 1948 (3rd edition). 414 p.

(370) *Ibid.* (p. 151–54).

(371) *Ibid.* (p. 197).

(372) *Ibid.* (p. 312).

(373) *Ibid.* (p. 314).

(374) *Ibid.* (p. 316).

(375) VIZE, V. YU. *Na "Sibiryakove" v Tikhiy okean* [*In the "Sibiryakov" to the Pacific Ocean*]. Leningrad, Izdatel'stvo Glavsevmorputi, 1934. 147 p.

(376) *Ibid.* (p. 142).

(377) VIZE, V. YU. *Na zemlyu Frantsa-Iosifa* [*To Zemlya Frantsa-Iosifa (Franz Josef Land)*]. Moscow, Leningrad, Zemlya i Fabrika, 1930. 176 p.

(378) VIZE, V. YU. "Reys ledokola *Malygin* na Zemlyu Frantsa-Iosifa v 1931 godu" [Voyage of the icebreaker *Malygin* to Zemlya Frantsa-Iosifa (Franz Josef Land) in 1931], *Trudy Ark.Inst.* Tom 6, 1933, p. 1–41. (Reference to p. 1–12.)

(379) VIZE, V. YU. "Trading navigation in the Kara Sea", *Polar Record*, Cambridge, No. 6, 1933, p. 90–94.

(380) VIZE, V. YU. "Tret'ya vysokoshirotnaya ekspeditsiya na *Sadko* 1937 goda" [The third high-latitude expedition in the *Sadko*, in 1937], *Prob.Ark.* No. 1, 1938, p. 73–77.

(381) VIZE, V. YU. *Vladivostok—Murmansk na "Litke"* [*Vladivostok to Murmansk in the "Litke"*]. Moscow, Izdatel'stvo Glavsevmorputi, 1936. 156 p.

(382) VIZE, V. YU., ed. *Ekspeditsiya na samolete "SSSR-N-169" v rayon "polyusa nedostupnosti". Nauchnyye rezul'taty* [*Expedition in the aircraft "SSSR-N-169" to the region of the "Pole of inaccessibility". Scientific results*]. Moscow, Leningrad, Izdatel'stvo Glavsevmorputi, 1946. 199 p.

(383) VOLKOV, N. A. "O ledovoy sluzhbe na vostochnoy trasse sevmorputi" [On the ice service in the eastern course of the Northern Sea Route], *Sov.Ark.* No. 5, 1941, p. 14–17.

(384) VOYEVODIN, N. "Morskoy put' v Sibir'" [Sea route to Siberia], *Sov.Sev.* No. 3, 1930, p. 62–83.

(385) *Ibid.* (p. 70).

(386) *Ibid.* (p. 74).

(387) *Ibid.* (p. 75).

(388) *Ibid.* (p. 75–76).

(389) VOYEVODIN, N. "Severnyy morskoy put'" [Northern Sea Route], *Sov.Aziya*, No. 3/4, 1930, p. 101–08.

(390) "Vozvrashcheniye sudov severovostochnoy ekspeditsii 1932 goda" [The return of the ships of the north-eastern expedition of 1932], *Byull.Ark.Inst.* No. 9/10, 1933, p. 279–83.

(391) VYSHNEPOL'SKIY, S. "Prioritet russkikh v stroitel'stve ledokolov" [The Russians were the first to build icebreakers], *Mor.Flot*, No. 2, 1948, p. 45–46.

(392) YANSON, N. M. "Plan raboty Glavsevmorputi v 1937 godu" [Plan of work for Glavsevmorput' in 1937], *Sov.Ark.* No. 2, 1937, p. 14–23. (Reference to p. 14.)

(393) *Ibid.* (p. 14–15).

(394) *Ibid.* (p. 17).

(395) YEREMENKO, A. S. "Zheleznodorozhnaya magistral' v polyarnoy tundre" [Railway line in the polar tundra], *Byull.Ark.Inst.* No. 6/7, 1933, p. 167–69.

(396) YEREMEYEV, N. A. "Obzor morskikh operatsiy v zapadnoy sektore" [Survey of sea operations in the western sector], in ZUBOV *et al.*, No. 405 below, p. 11–37.

(397) YEREMEYEV, N. A. "Vo vseoruzhii podgotovit'sya k arkticheskoy navigatsii" [Let us use all the means at our disposal to prepare for the arctic navigation season], *Sov.Ark.* No. 4, 1938, p. 7–12.

(398) YEVGENOV, N. I. "Samolet na sluzhbe severnogo morskogo puti" [The aircraft in the service of the Northern Sea Route], in ANVEL'T, YA. YA., ed., and others, *Vozdushnyye puti severa* [*Air routes of the north*], Moscow, Izdatel'stvo Sovetskaya Aziya, 1933, p. 139–66.

(399) YEVGENOV, N. I. "Severovostochnaya ekspeditsiya 1932 goda" [The north-eastern expedition of 1932], *Byull.Ark.Inst.* No. 5, 1933, p. 118–24.

(400) YEVGENOV, N. I. "Vysokoshirotnaya ekspeditsiya na 1/p *Sadko*" [High-latitude expedition in the icebreaking ship *Sadko*], *Byull.Ark.Inst.* No. 10, 1935, p. 322–28.

(401) ZHDANOVA, N. T. "Vrednaya vylazka B. V. Lavrova v Moskovskom dome uchenykh" [B. V. Lavrov's harmful sally at the Moscow House of Scientists], *Sov.Ark.* No. 9, 1937, p. 9.

(402) ZINGER, M. "Upushchennyye vozmozhnosti" [Lost opportunities], *Sov.Ark.* No. 1, 1938, p. 32–36.

(403) ZUBOV, N. N. *L'dy arktiki* [Ice of the Arctic]. Moscow, Izdatel'stvo Glavsevmorputi, 1945. 360 p. (Reference to p. 346–48.)

(404) *Ibid.* (p. 350).

(405) ZUBOV, N. N., ed., and others. *Arkticheskiye navigatsii. Sbornik pervyy* [*Arctic navigation seasons. First handbook*]. Moscow, Leningrad, Izdatel'stvo Glavsevmorputi, 1941. 268 p.

(406) *Ibid.* (p. 175).

INDEX

SVALBARD

ZEMLYA FRANTSA IOSIFA
(Franz Josef Land)

BARENTS SEA
(Barentsovo More)

Severnaya
Zemlya

M Molotova

Vardö
Murmansk
KOL'SKIY
P-OV

Novaya

Zemlya

M Zhelaniya

O.Uyedineniya

Proliv Shokal'skogo

Proliv Vil'kitskogo

O.Tyrtova

O.Ko

White Sea
(Beloye More)

Arkhangel'sk

Matochkin Shar

KARA SEA
(Karskoye More)

TAYMYR

Nord

Karskiye Vorota
O.Vaygach
Yugorskiy Shar

O.Diksong

Amderma

Pyasina

Pechora

Novyy
Port

Us¢-Port

Dudinka

Noril'sk

Dudypta

Volochanka
Kheta
Khatanga

SIY

U R A L

TYUMEN'SKAYA

OBLAST'

Igarka

Kotuy

KRAY

Nizhnyaya Tunguska

60°

Ob'

N

Yenisey

Podkamennaya
Tunguska

K H R E B E T

Tyumen

Zlatoust

Tobol'sk

Tobol

Omsk

Yeniseysk

Maklakovo
Verkhnyaya
Tunguska

Pridivnensk

KRASNOYARS

60°

Irtysh

Novosibirsk

Krasnoyarsk

Kach

Irkutsk

Kilometres
0   200   400   600   800
0   100   200   300   400   500
Statute Miles

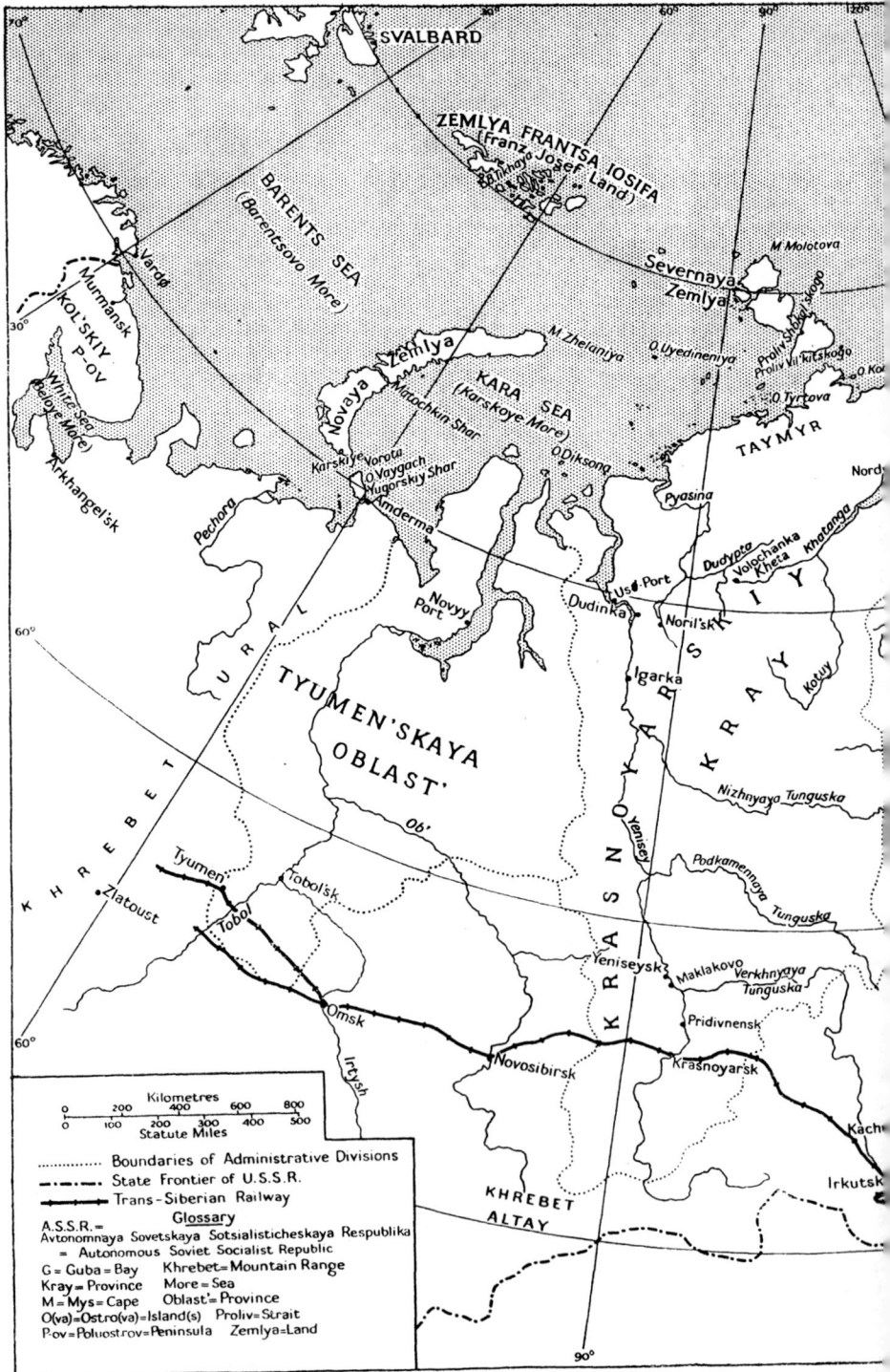

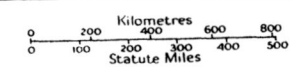

.... Boundaries of Administrative Divisions
—·—·—·— State Frontier of U.S.S.R.
━━━━━ Trans-Siberian Railway

KHREBET
ALTAY

Glossary
A.S.S.R.=
Avtonomnaya Sovetskaya Sotsialisticheskaya Respublika
   = Autonomous Soviet Socialist Republic
G = Guba = Bay   Khrebet=Mountain Range
Kray = Province   More = Sea
M = Mys = Cape   Oblast'= Province
O(va)=Ostro(va)=Island(s)   Proliv=Strait
P.ov=Poluostrov=Peninsula   Zemlya=Land

90°

Map 9.

150°

180° 80°

O.Vrangelya

CHUKCHI SEA

BERING STRAIT

Uelen

70°

Proliv Longa (Chukotskoye More)

Kolyuchinskaya G.

EAST SIBERIAN SEA

M.Shmidta

CHUKOTKA

Za'Providentiya

(Vostochnosibirskoye More)

M.Shelagskiy

Pevek

BERING SEA

O.Bennetta

(Beringovo More)

Chaunskaya G.

160

Novosibirskiye O-va

M.Chetyrekhstolbovoy

Anadyr

Anadyr

Ioy Pravdy

Proliv Sannikova

Ambarchik

K

160

LAPTEV SEA

M.Shalaurova

Nizhnekolymsk

R

(More Laptevykh)

Proliv Lapteva

Russkoye

Ust'Ye

A

evnikova

Saskylakh

Tiksi

Kazach'ye

Kolyma

Zyryanka

Y

KAMCHATKA

Anabar

Bulun

Yana

Indigirka

Olhonoina

Magadan

Olenek

Verkhoyansk

Ege-Khaya

Moma

S

K

YAKUTSKAYA

A.S.S.R.

Lena

Sangar-Khaya

V

OKHOTSK SEA

Petropavlovsk-na-Kamchatke

Vilyuy

Yakutsk

Aldan

(Okhotskoye More)

O

50

Peleduy

R

Vitim

A

B

150

Baykal

A

H

K

Vladivostok

120°

40

Northern Sea Route.

Lightning Source UK Ltd.
Milton Keynes UK
UKOW04f0709080614

233040UK00001B/36/P